EARTHING & BONDING IN HAZARDOUS AREAS

Ian Staff

ELECTRICAL TRAINING CONSULTANT

First Edition published 2021

2QT Limited (Publishing)

Settle

North Yorkshire

BD24 9BZ

Copyright © Ian Staff 2021

The right of Ian Staff to be identified as the author of this work has been asserted by him in accordance with the Copyright, Designs and Patents Act 1988

All rights reserved. This book is sold subject to the condition that no part of this book is to be reproduced, in any shape or form. Or by way of trade, stored in a retrieval system or transmitted in any form or by any means, electronic, mechanical, photocopying, recording, be lent, re-sold, hired out or otherwise circulated in any form of binding or cover other than that in which it is published and without a similar condition, including this condition being imposed on the subsequent purchaser, without prior permission of the copyright holder

Cover design: Dale Rennard

Images supplied by author

Printed in the UK by Lightning Source UK Limited

ISBN 978-1-914083-11-2

About the Author

I am an Electrical Training Consultant and carry out Electrical Training for a company in Hull by the name of Humberside Offshore Training Association Ltd. (H.O.T.A.), Malmo Road. Before my 15 or so years at H.O.T.A. as a Trainer/Assessor I was 38 years with BP, seven of those years as their Instrument/Electrical Training Officer in charge of all Instrument and Electrical Training in their Training Department where I obtained my Training and Assessing Qualifications.

Following on from my last books 'Hazardous Areas for Technicians', 'Inspections in Hazardous Areas' and 'Motors in Hazardous Areas' I have produced this book 'Earthing and Bonding in Hazardous Areas' to assist Technicians who work in hazardous areas. Earthing and Bonding in Hazardous Areas is again aimed at Technician level. Earthing only becomes a problem when something goes wrong or when tests have to be carried out and it is good to understand what systems and methods are in place.

Ian Staff SB St J. Electrical Training Consultant.

Recommended Publications Regulations & Standards for Hazardous Areas:

ISBN 978-1912014958

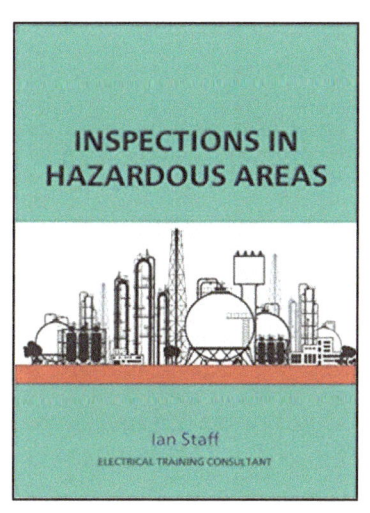
ISBN: 978-1913071615

ISBN: 978-1914083013

1 – BS7671: IEE Wiring Regulations — 978-1785611704

2 – EEMUA: Handbook-Explosive Atmospheres — 978-0859312127

3 – IEC 60079: Standard-Explosive Atmospheres — INTERNET-COST!

4 – ISO 80079: Non-Electrical/Mechanical — INTERNET-COST!

5 – DSEAR: Regulations-Explosive Atmospheres — 978-0717666164

6 – 1989 Electricity at Work Regulations — 978-0717666362

Humberside Offshore Training Association (HOTA):

HOTA was established in 1987 and based in Hull, East Yorkshire and is a Quality Training Provider offering nationally approved training and tailor made bespoke courses to meet individual and company specific training needs.

A Limited Company with Charity Status, all surplus funds generated are invested back into enhancing training and delegate facilities at The Cullen and The Ellis Buildings – Malmo Road and its Albert Dock Site.

HOTA is renowned for its flexibility, professionalism and its industry experienced, highly trained team of trainers delivering courses when, where and how they are required.

Since 1999, HOTA has been offering the JTL Approved Electrical Equipment in Hazardous Areas (CompEx) Course and over the years has continually added to the existing Training Portfolio. HOTA now offers the CompEx Main Course (EX01–EX04), CompEx Refresher (EX01R–EX04R), CompEx Foundation (EXF), CompEx Dust (EX05–EX06) and CompEx Mechanical Equipment in Hazardous Areas (EX11).

HOTA's assessment suite is a world class facility and delegates are allotted their own fully equipped assessment bay. All courses are conducted by highly experienced JTL approved trainers and assessors.

To complement the CompEx training courses, HOTA conducts various City & Guilds Approved Electrical Courses as well as bespoke training courses including the one day Introduction to Hazardous Areas and pre CompEx Glanding Practice.

HOTA provides Nationally Approved and Bespoke Electrical and Mechanical Training Courses including:

JTL Approved:

- CompEx EX01 – EX04 Electrical Equipment in Hazardous Areas and Explosive Atmospheres (5 Days)
- CompEx EX01R – EX04R Refresher after 5 Years (3 Days)
- CompEx EX05 – EX06 Electrical Equipment in Dust Atmospheres (3 Days)
- CompEx EX11 – Mechanical Equipment in Hazardous Areas & Atex (3 Days)
- CompEx EXF – Foundation Course (2 Days)

City & Guilds Approved:

- IEE 18th Edition 2382-2-18 Level 3 Wiring Regulations
- PATS 2377-22 Level 3 Award In-Service Inspection & Testing of Electrical Equipment
- 2391-02 Level 3 Inspection & Testing
- City & Guilds 2919-01 Level 3 Award in Domestic, Commercial and Industrial Electric Vehicle Charging Equipment Installation

Bespoke Training:

- Introduction to Hazardous Areas
- Pre-CompEx Glanding Course
- Mechanical Joint Integrity
- Atex Familiarisation
- Electricity at Work Regulations – Practitioners and Non-Practitioners
- Static Electricity in Hazardous Areas.

HOTA offers more than 100 nationally approved training and tailor-made bespoke courses across a number of sectors including:

Offshore Oil & Gas Training Courses

Maritime Training Courses

Wind Turbine and Renewable Energy Training Courses

Emergency Rescue Response Vessel Training Courses

CompEx Electrical and Mechanical Training Courses

City & Guilds Training Courses

Medical – First Aid Training Courses

Health & Safety Training Courses

Firefighting Training Courses

Specialist Training Courses

Royal Yachting Association Training Courses

For a full list of courses and dates please visit HOTA's facilities:

Website: www.hota.org

Telephone: +44(0) 1482 820567

Email: bookings@hota.org

Address: Malmo Road, Sutton Fields Industrial Estate, Hull. HU7 0YF

HOTA's purpose-built training facilities have ample free onsite parking and free WI-FI throughout the site.

Delegates can also enjoy a free two course lunch and refreshments in the onsite 100-seater restaurant.

Introduction

This book: 'Earthing and Bonding in Hazardous Areas' is the fourth book in the series. My first book 'Hazardous Areas for Technicians' is very successful, explaining what Atex is and how Atex equipment is installed and maintained in Hazardous Areas as well as things like 'Cathodic Protection' etc. My second book 'Inspections in Hazardous Areas' is mainly for Companies who carry out their own Inspections of Hazardous Area Electrical, Instrument and Mechanical Equipment and focuses how Electrical, Instrument and Mechanical Inspectors go about understanding standards, why the equipment is inspected and the terminology. My third book Motors and Control in Hazardous Areas is all about the History of the electric motor ans all of the different motors that Electrical Technicians may come across in their Hazardous Areas from DC to AC.

I have always said that how can anyone go about maintaining or finding faults on equipment and systems in Hazardous Areas if they do not know or understand how that equipment or systems works to start with! 'Earthing in Hazardous Areas' starts off with the history of the colours and symbols right up to how to carry out the installation and testing and why various tests are done.

The book explains the importance of the Earthing and all of the different methods that can be installed into the ground to achieve this. The important of the 'Star Point Earth' of the Distribution Transformer and that most earthing circuitry leads back to there is discussed along with plant Lightning Protection.

Earths on Variable Frequency Drives can be a huge problem for companies with them. The book goes on to discuss how these High Frequency Earthing Systems can destroy bearings in the motor and the bearings in anything that is being driven such as a pump or compressor and how to maintain the Earthing Systems to see if they are still working efficiently and if there are any problems forming.

The book also looks in detail at Bonding of Vehicles which come onto a Chemical Factory and Equipment such as vessels & pipework which are part and why this is so important and also what the difference is between Earthing and Bonding and why each of these topics are so important in a Hazardous Area.

There are static sparking dangers involved in bonding equipment of companies that come onto the plant to carry out cleaning and cutting using very high pressure water jetting and towards the end of the book we look at with the problems involved with Static Electricity on both Plant fixed Equipment and on Humans that work on the plant every day and what steps a company will take to prevent this including the anti-static devices that can be used in a workshop.

Finally all the different forms and documentation is discussed and examples of the different forms are shown and how Technicians may complete them!

Contents

About the Author 3
Humberside Offshore Training Association (HOTA): 4
Introduction 6
Contents 7

Earthing: Standard for Graphical Symbols: IEC 60417 10

History of Earth Colours: 11
Standard for Earthing Colours: BS7671 (18th Edition) 11
Distribution Transformer: 12
Three Phase Distribution Transformer 13
Getting an Electric Shock: 14
Fault to Equipment Metal Case: 15
Earthing Systems: TNC-S/TNC/TN-S/TT/IT 17
    Earthing System TNC-S: 18
    Earthing System TNC: 20
    Earthing System TN-S: 21
    Earthing System TT: 22
    Earthing System IT: 23
Earthing Methods: Rods/Plates/Lattice/Electrolyte/Electrodes/Coils: 24
    Standard Earth Rods: 25
    Earthing Plates: 27
    Earth Lattice: 28
    Electrolytic Earth Electrode: 29
    Earth Electrodes in Concrete: 30
    Earth Coil Electrodes: 31
Earth Rods and Accessories: 32
Joining Earth Electrodes: 33
Earth Pits: 34
Earthing Conductors: 35
Earthing Electrode Backfill Compounds 36
    Marconite: 36
    Bentonite: 36
    Gel: 36
Chemical/Gel Earth Rods: 37
Soil Resistance Earth Testing: 38
Soil Resistivity Testing: 39
Ground Resistance Readings: 40
Testing Earth Electrode Resistance: 41
Earth Clamp Meter: 42
Earthing Drawing Documentation & Drawings: 43
Earth Loop Impedance TN-S: 46

Earth Loop Impedance TT:	47
Testing Motor Earth Path:	48
Earth Leakage Units and RCDs:	49
Residual Current Circuit Breaker (RCCB):	50
Earth Leakage Unit (ELU):	51
Voltage-Operated Earth Leakage Unit:	52
Motor Earth Leakage Protection:	53
Clean Earths:	54
Barrier Box Clean Earth Bar:	55
Barrier Box Clean Earth Bar:	56
Clean Earth Intrinsic Safety:	57
Lightning Protection:	58
Lightning Protection Standards: IEC 62305, IEC 62561, EN 50164	59
Lightning Protection Zones:	60
Lightning Ring Earthing Conductors:	61
Voltage Surge Protection Devices (SPDs):	62
High Voltage Earthing:	63
Single Earth Panel Earthing:	64
Bus-Zone Protection:	65
Restricted Earth Fault:	66
Earthing Transformer & Delta Earth Protection:	67
High Voltage Busbar Earth:	68
High Voltage Circuit Earth:	69
Transformer Earthing:	70
Distribution Transformer Earthing:	71
Centre Tap Transformer Earthing:	72
Instrument Supply Transformer Earthing:	73
Floating Transformer Earthing:	74
Step Up Transformer Earthing:	75
Auto Transformer Earthing:	76
Current Transformer Earthing:	77
Generator Earthing:	78
Generator Earthing 1:	79
Generator Earthing 2:	80
Large Portable Generators:	81
Chemical Storage Tank Earthing:	82
Earthing Chemical Storage Tanks:	83
Earthing and Bonding Floating Roof Tanks:	84
Variable Frequency Drive Earthing:	85
Stator to Rotor Coupling Current:	86
Rotor to Shaft Current:	87
Stator to Shaft Current:	87

- Stator to Ground Current: 88
  - Fluting: 88
  - Frosting (Pitting): 88
- Electrical Earthing Drawing Symbols: 89
- Double Insulation: 90
- Water Jet Cleaning and Cutting: 92
- Steaming Out Bay: 93
- Socket Polarity and Earth Tester: 94

Bonding: 95
- Earthing and Bonding? 96
- Equipotential Bonding: 97
- Exe Increased Safety Junction Boxes: 98
- Exd Flameproof Junction Box: 99
- Bonding Flexible Hoses: 100
- Bonding/Earthing of Large Vacuum Units: 101
- Bonding/Earthing of Road Tankers: 102
- Bonding/Earthing of Road Tankers Earth Monitoring: 103
- Bonding of Pipe Flanges: 104
- Bonding of Pipes: 105
- Earthing of Pressure Vessels: 106
- Filling Cans with Hazardous Liquid: 107
- Bonding of Instrument Junction Box: 108
- What is Static Electricity? 109
- Antistatic Footwear: 110
- Antistatic Wrist Straps: 111
- Antistatic Mats: 112
- Antistatic Spray: 113
- Protection: 114
- Hazardous Areas: 116
  - Zones, Categories & Equipment Protection Levels (EPLS): 116
  - Gas Groups: 117
  - Dust Groups: 118
  - Temperature Classification: 119
- Several Earthing Terminologies: 120

INDEX 121

Earthing:
Standard for Graphical Symbols: IEC 60417

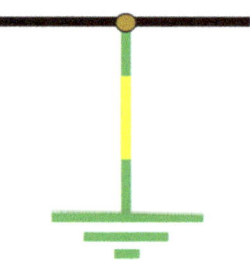

Earthing in our Hazardous Areas is mainly taken for granted and it is not until a routine maintenance task comes up where testing has to be done that people realise what types of earthing there are. In my opinion there are four main reasons for earthing, namely for:

1) **Personnel Protection** – Metal items are earthed to take fault current and metalwork is equipotential bonded.
2) **Equipment Protection** – High fault currents for long periods of time cause very severe damage.
3) **Lightning Protection** – Protects the plant against strikes, but if it does get struck it provides a path to earth to reduce damage.
4) **Antistatic Purposes** – Static electricity is a problem. It can cause sparking.

If you ask what is the difference between earthing and bonding many people will say '**not a lot**' and in some circumstances they may be right. Just look at it as: earthing involves connection to the ground, and bonding is usually, say, metal to metal. Because the earth is a conductor of electricity, this does provide a path back to the **'distribution transformer star point'** where the fault electrons want to go.

Earthing is essential for the safety primarily of personnel, but also plant, and without earthing we may have more electrical accidents. It is essential that all metalwork on a plant is around the same potential otherwise there is an opportunity for sparking or electric shock. Also we are looking at earth electrode **surface area** rather than depth as the more surface area there is the faster we can dissipate earth fault currents.

The typical symbol for earthing is the one above, with **three** lines forming the arrowhead. Drawing symbols would be black and maybe not so elaborately coloured as my example above although yellow and green are the colours for earthing cable, sleeving and tape. Most of the symbols that you will encounter are shown later in the book.

This earthing section shows the different Systems i.e. **TNC – TNC-S – TN-S – TT – IT** and also the different methods of earthing i.e. Rods, plates, lattices etc. It also deals with how and why we earth generators and transformers. High voltage systems are touched on as far as Technician level is concerned. Different methods of testing are also discussed with easy-to-understand diagrams of how to carry out the Earth Tests and how to use the different test instruments. In America Earthing may be called **Grounding**.

History of Earth Colours:
Standard for Earthing Colours: BS7671 (18th Edition)

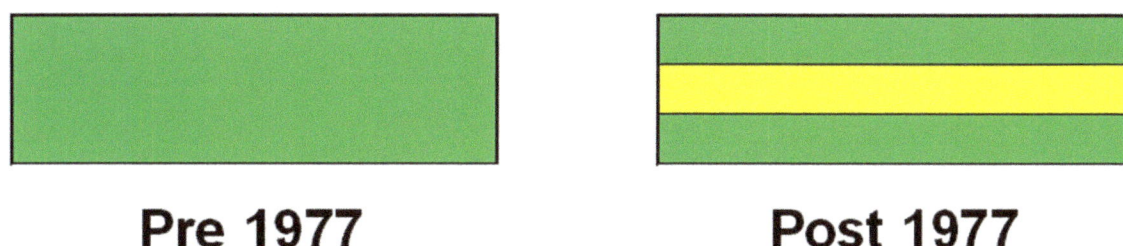

Pre 1977   **Post 1977**

Electrical earthing wiring, sleeving and tape is green and yellow these days. When did it change? Many older Technicians like myself remember when the earth wire in an electrical circuit was just pure green.

Then after **1977** this changed and electrical earthing wires, tape and sleeving became green and yellow. Wiring colours changed in **2004** to harmonise with Europe, but in the UK earth colour, green and yellow remained the same.

Other countries may differ from our green and yellow colour. In Canada, North America, Australia, India and New Zealand the earth colour may be green and yellow or like our pre 1977, just green.

Screen Colour 1   **Screen Colour 2**

When looking at Intrinsic Safety system multicores they have 'Screens' in the cables which are metal foils, which look like silver paper, wrapped around the multicore pairs. The foil is not, of course, silver paper as it is a conductor on the inside and an insulator on the outside.

The idea is that each pair in the multicore is in a foil Faraday Cage and then another foil screen around all of the pairs called an 'Overall' screen. These screens stop outside invading voltages being induced into the IS system.

In contact with the internal conductive part of the screen throughout the cable is a bare wire which, when the cable cores are stripped and exposed, has to have an insulated sleeve fitted. Many Technicians actually call the bare wire the 'Screen'.

The screen is connected as per Hook up/Loop diagram to clean or dirty earth hence earthing the inner conductive part of the silver foil screen. The sleeve colour for the screen is, as per company policy, commonly green or white.

1) Do you use the above Intrinsic Safety Screen Colours on your plant?
2) If the answer to '1' is 'No' then what colour do you use?

Distribution Transformer:

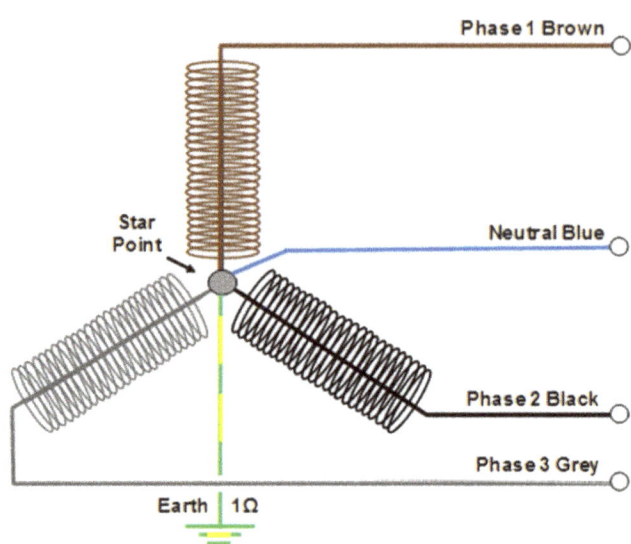

The Distribution Transformer will **usually** be a 'Star' connected secondary winding. It is here where the main earth and the neutral for a particular wiring system is connected. On the following page we discuss why the star point is earthed in the first place and in what circumstances there may not be one, or a very high impedance one **(IT earthing system)**. Should the distribution transformer be a 'Delta' connected secondary then an extra transformer may have to be added called an **'Earthing Transformer'** which would have a star point. This would, of course, be a very expensive way of equipment design and later we will look at how this transformer may be connected into the system.

The star point of the Distribution Transformer, in the diagram above, should have an impedance of **1Ω** or better. May I just add here that 'better' may be very difficult to achieve. The low impedance ensures a very low resistance path back should there be any fault currents. Later on we will discuss how it will be possible to get a rod resistance of **1Ω**.

The transformer in the diagram above can be connected in several different ways as far as the earthing systems go. It can be TN-C, which of course is **dangerous in hazardous areas** TT or TN-S. All of these earthing systems along with the IT earthing system are discussed later.

The voltage between each of the phases in the above diagram is around **415V**. Some years ago this may have been **440V**. The voltage between any of the 3 phases and neutral is around **230V**. Again, some years ago this would have been **250V**. The voltage between any of the 3 phases and earth is around **230V** and again, some years ago this would have been **250V**. Primary voltages in the past were 6.6KV and the resultant voltage was 440V and 250V. These days the voltage may be 11KV and 3.3KV so the resulting voltage would be less.

1) Do you know the location of your Distribution Transformer(s)?

2) If 'YES' can you identify the Star Point Earth?

3) What voltage is the 'Secondary' of your Distribution Transformer?

4) What voltage is the 'Primary' of your Distribution Transformer?

Three Phase Distribution Transformer

Firstly let us have a look at a typical distribution transformer which supplies your Hazardous Areas via substations (HV) and switch rooms (MV). You will notice that there are 3 phases, a neutral and an earth. This transformer is **usually** connected in what is known as 'Star' as discussed on the previous page, and has a point where one end of the three windings are connected together called, as you would imagine, a **'Star Point'** from where we also obtain our Neutral.

For safety reasons this 'Star Point' is earthed to the ground with an earth reading of **1Ω** or better and this earth (and hopefully not humans!) forms the return path for all faulty equipment, through the ground and back to the distribution transformer.

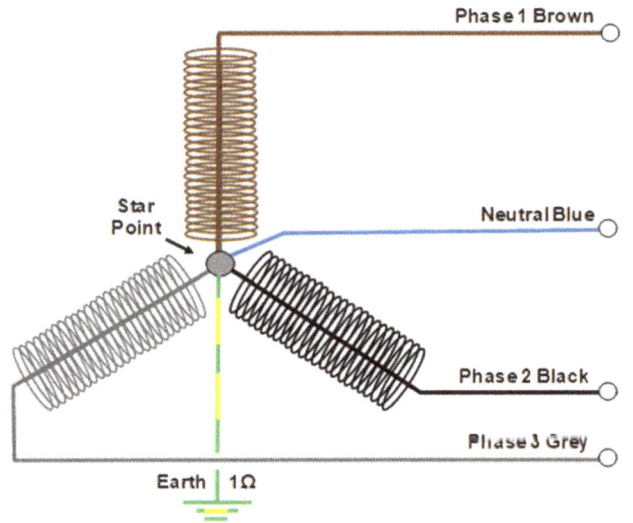

So looking at the diagram on the left we can see the 3 phases which these days are brown, black and grey (replacing red, yellow and blue).

Back at the transformer on the left, neutral and earth are exactly the same potential, but out in the field the voltage between earth and neutral can differ significantly. I have in the old days measured a voltage of 5V between earth & neutral which could of course cause sparking which is not intrinsically safe.

One question that is regularly asked is, if the safety earth is 230V between it and a phase, the same as the neutral, and is the cause of many electrocutions, why not remove it? The answer is **WE CAN**, and almost do with the **'IT'** Earthing System which I will explain later. The only problem with removing it is that many protection systems which guard faulty equipment rely on that earth reference being there.

Let us just imagine that the star point earth is removed on the distribution transformer. You might say now that I will never get another potentially fatal shock to earth and you would be correct, **PROVIDING** that the transformer neutral, which also emanates from the star point, **NEVER** got a fault to earth. If this happened then suddenly there would be an earth reference and a 230V potential from live to earth, and because you were not expecting this you may have no metal connected to earth designed into the circuit, making faults to the case of equipment lethal. This could happen if, say, the telephone company hammered a spike into the ground without scanning, penetrated the cable and went through the neutral without the other phases.

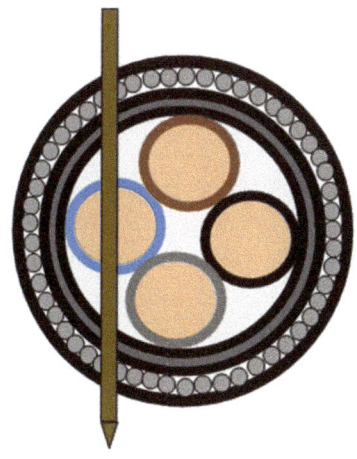

I appreciate that this might be picked up on some protection relays such as a Solkor Relay, but they usually come into play if the spike was to go through the phase conductor.

Getting an Electric Shock:

Let us discuss how people get electric shocks off equipment and how we try to make all metalwork not connected to the electrical system the same potential **(Equipotential Bonding)** so people do not receive shocks when they go from touching the metalwork of an appliance **(Exposed Conductive Part)** during a fault to, say, touching a water pipe **(Extraneous Conductive Part)**, because all metalwork is at the same potential.

It only takes around 33mA (33 thousandths of 1 Amp) to kill Mr or Ms Average so as you can appreciate much higher currents can be around under fault conditions.

Well, on a chemical factory there is obviously a lot of metal to be bonded, but let me use two examples of Cable Tray & Ladder Rack. These are what are called **'Cable Management'** and is the way cables are run through the plant. Cables and electrical equipment can be fixed to the metalwork of the tray, but it is not a direct part of the earthing system and is not meant to carry fault current, so at some point would have a **Bond** of its own. The cable tray would be called an **'Extraneous Conductive Part'** and would be connected to the plant earthing system as part of the equipotential bonding along with the structure itself and all of the columns, tanks and vessels.

The above cable tray would be an ideal **Equipotential Bonding** situation. So if we fixed a flameproof Exd metal socket outlet or switch to the cable tray we would run a 4mm **bonding wire** from the metal socket or switch case to the metal work of the cable tray just as you would, in a house, take a bonding wire to all pipework. If the switch is increased safety then we cannot fix a bonding screw to a plastic case so we put an earth tag on one of the glands and connect a 4mm bonding wire from this to the cable tray.

The **Exposed Conductive Part** could become live if it was not earthed in the event of a fault to earth and the **Extraneous Conductive Part** could be other metalwork you could touch to receive an electric shock. If we **Bonded** the two together **there would be no difference in potential**. Remember that in our Hazardous Area we are not just concerned with human protection, but with sparking as well. Two different items i.e. metal socket outlet and cable tray could spark momentarily in the event of a fault current to earth if they did not have an **Equipotential Bond**.

1) Is Electric Shock Treatment Procedures up in all Substations/Switch Rooms?

2) Are all Technicians trained in Electric Shock First Aid?

Fault to Equipment Metal Case:

Basically speaking equipment is earthed to protect life and equipment and the mere fact that the ground on which we stand on conducts electricity does assist that process by providing a path back to the distribution transformer earth, but can also sometimes be the cause of the electric shock. People ask 'if the transformer star point was removed nobody would receive a shock from live to earth so why do they not remove it?' We will talk about this on the next page and discuss the safety aspects as to why the star point earth is in place.

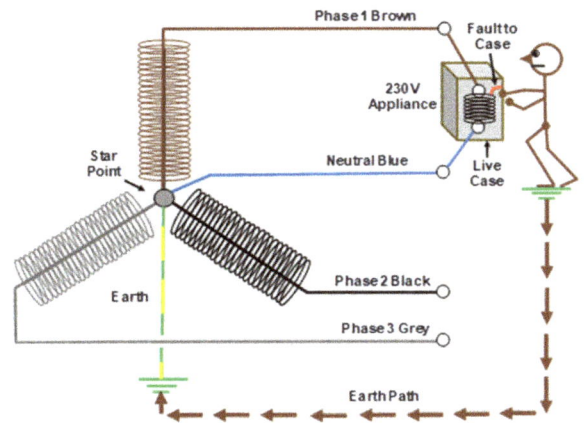

Let us take the example in the diagram to the left, which is a 230V appliance with a metal case. There is a fault on from the brown phase. This makes the metal case part of the phase and we would call this 'live'. You will notice that the case, in the diagram left, is not earthed so therefore it would just sit there, probably working satisfactorily as the electrons would be travelling down the phase through the load and back to the star point through the neutral. This then would be called an **'Exposed Conductive Part'**. (Part of the system, but not meant to be live.)

That is until someone touches the metal case and connects it to earth by the mere fact that they are standing on it. Some of the electrons are still returning via the neutral, but some are going through the person, down through the ground and up the star point earth. The fuses will not blow because there is not sufficient low resistance according to our friend Georg Ohm. If more than around 30mA (30 thousandths of 1 amp) in current was to flow through an average human then this shock could be fatal especially if the electric current path was across the heart. The severity of the shock also depends upon several factors of voltage, current & resistance:

1) How high was the voltage? (Above that would be 230V)

2) How moist was the person's skin and resistance of the body?

3) How good a contact did they have with the live metal case?

4) Were they standing on damp earth, wood, carpet etc? Ground resistance?

5) Did their shoes have rubber soles i.e. Trainers etc.? Sole resistance comes in here.

6) Was the shock across the heart? (Not exactly earthing, but more fatalities.)

7) How long were they in contact with the voltage?

8) Was the appliance Residual Current Device (RCD) or Earth Leakage protected?

If the above lists are hazards that might come from an electrical system then we must have precautions, equipment and test procedures in place to prevent them. As we proceed through the book we will look at the different earthing systems that are available and how we protect against electric shock and equipment damage both in domestic and industry settings.

Let us take a very similar situation to the one on the previous page, except this time we will earth the appliance case as below, either by driving a rod into the ground or providing a physical hard wire back to the star point earth. So now we can say that our **'Exposed Conductive Part'** is earthed. These would be 'Class I' Appliances.

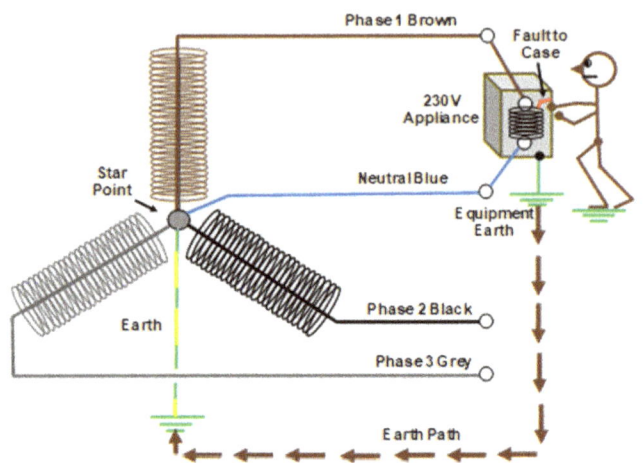

Let us just change the situation slightly. Our stick person is in contact with the metal case **(Exposed Conductive Part)** at the moment there is a fault from phase to earth. The earth path would be through the 'point of least resistance' which would, **hopefully if it was tested correctly,** be the equipment earth, so if the appliance was earthed properly the fuse would blow. With this example, the person may still get a tingle but a fatal shock now is very unlikely.

This relies on the equipment having a good earth in the first place, which is why PAT testing, earth path testing, insulation testing and earth loop impedance, earth rod value checks etc. are performed by competent people who can both do the tests correctly and interpret the results. **The person must not be a better earth.**

As mentioned earlier, as we go through the book we will look at the various earthing systems and see which are suitable for our Hazardous Area. We will look at various earth protection equipment that can be added on or included.

Again, if the equipment had earth leakage protection **(ELU)** or was protected by a residual current device **(RCD)** then it is likely that this may operate. Usually on a direct dead short to earth the fuses may go first. Remember that 'earth leakage' which is described in future pages is exactly what it says, 'leakage', not dead short circuits.

Therefore this action of providing the case of the equipment with a good, tested earth will protect personnel if there is a case fault to earth.

The equipment metal case (Exposed Conductive Part) earth will not:

1) Protect personnel if they get across live and neutral because they have not isolated the equipment that they are working on.

2) Protect personnel if they get across live to the equipment case because they have not isolated the equipment that they are working on.

3) Protect the equipment inside the enclosure from damage due to high fault currents, although other protection devices like fuses etc. may.

4) Protect the person if **THEY** are a better earth than the protective earth.

This is why a good earthing system, regularly tested to the correct resistance, is essential along with other measures such as correct size fuses and protection. I cannot stress enough that the personnel carrying out the testing must be competent and be able to interpret the test results and know immediately if something is wrong or dangerous.

Earthing Systems: TNC-S/TNC/TN-S/TT/IT

Now we are going to discuss the different earthing **'Systems'** and why the **TNC & TNC-S** Systems would be ok for domestic, but would be dangerous to have feeding electrical equipment in our Hazardous Area.

Faults with the TNC-S System include:

1) Consumer neutral/earth may rise in voltage and become a different potential to true earth. If this were to happen it could cause sparking.

2) Earthed metalwork becomes dangerous if neutral broken. This of course includes extension leads and feeds to external buildings.

3) Core balance units may fail to function under certain conditions.

TN-S & TT Systems are ideal for Hazardous Area locations and many chemical factories would have a mixture of the two.

TN-S: Earth is usually the **S**teel **W**ire **A**rmour of the supply cable.

TT: Terre Terre – earth is provided via a rod at the distribution transformer and another rod at the consumer.

I will also explain why the **IT** systems **(Isolated Terre)** may also not be an ideal choice in some circumstances, but on ships and oil platforms may be ideal as power loss could cause the ship to drift and maybe collide.

Oil platforms could become dangerous with no power at all. It may however not be ideal for your plant. When I have explained how it works later I will let you decide.

Meaning of the letters in the different systems:

T – Terre (Ground) direct connection of a point with Earth

N – Connection to Earth via Network

C – Combined Earth & Neutral

S – Separate Earth. (Not quite how it is worded!)

I – Isolated: Star point is connected to earth via High Impedance or not at all

In the next few pages we will look at each individual earthing system and discuss its suitability to be in our Hazardous Area.

1) What Earthing System is used on your plant? TNC   TNC-S   TN-S   TT   IT

2) Do you understand why a TNC-S System might be dangerous on your plant?

Earthing System TNC-S:

The next few pages demonstrate the different Earthing Systems that can be installed both in domestic and industrial situations. Let us first have a look at the TNC-S Earthing System which is used for domestic premises, but which I must iterate IS NOT SUITABLE FOR A HAZARDOUS AREA!

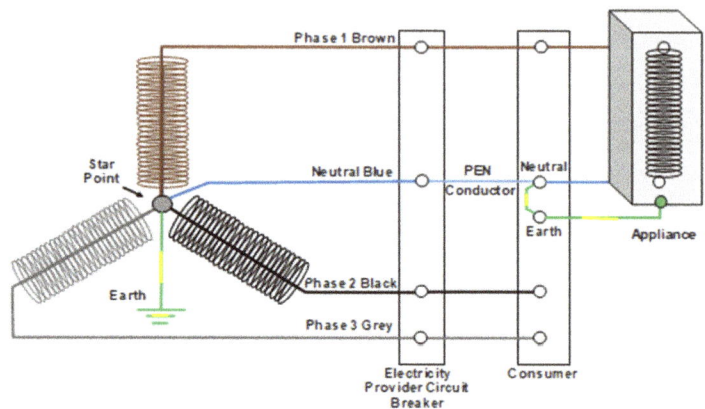

If we look at the diagram to the left we can see that the star point of the distribution transformer is earthed so this part of the system is **TN** because it is neutral connected to terre, '**C**' because on the route from the transformer to the consumer earth and neutral are combined and '**S**' because at the consumer the earth is a separate connection from the neutral.

Let us look at this TNC-S Earthing system in a little bit more detail. We have linked the Earth directly off the 'Neutral Block' at the consumer. The Neutral is now called a **P**rotected **E**arth & **N**eutral **(PEN)** conductor. I have shown an appliance plugged into phase and neutral and earthed to the linked earth connection. Next I have pointed out four problems with **TNC-S**.

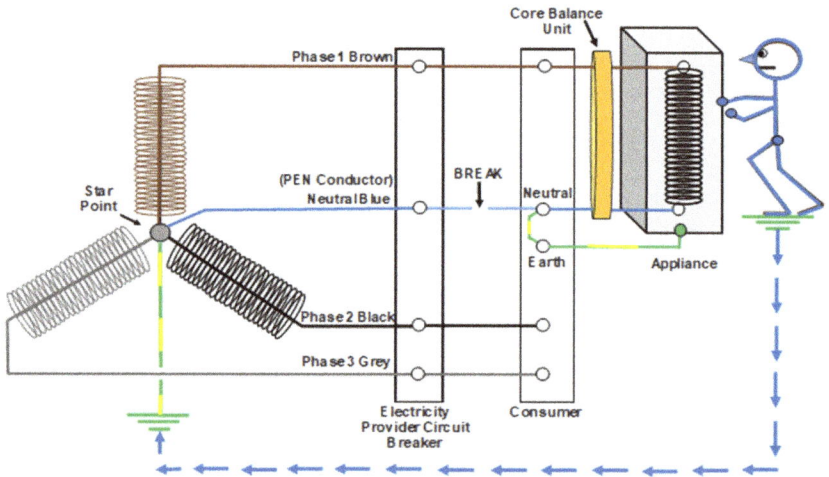

1 – If we now have a break (open circuit) in the Neutral between the electricity provider circuit breaker and the consumer **as above** we now have a VERY DANGEROUS situation. When switched on our appliance would not work because of the broken neutral so the electrons cannot get back to the star point of the transformer. Because the earth is linked to the neutral after the break, the case of the appliance would be acting as the neutral waiting for a path back. If someone (our stick person) was to touch the metal case while standing on earth then this person would become the neutral and provide the path back through them and the earth that they are standing on, along the ground and up the star point earth to the star point of the transformer and invariably would receive an electric shock which could be fatal.

2 – It is also dangerous in another way. I have drawn a Core Balance unit in yellow in the above diagram. This of course would not work and would not protect anyone as there is no fault to earth and the cores would still be balanced with the fault path.

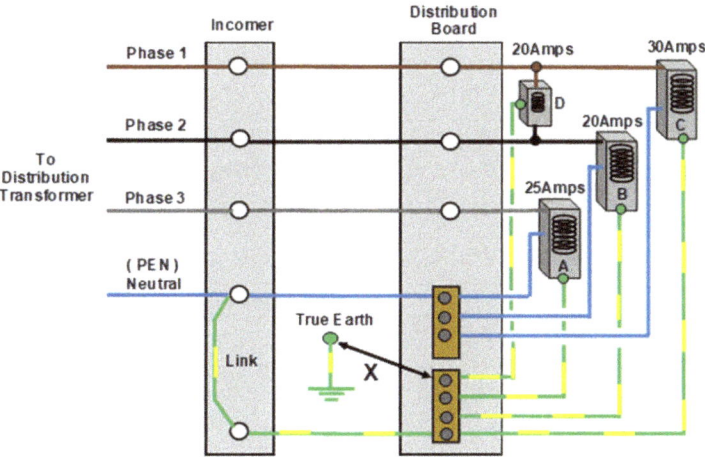

3 – Another problem with the TNC-S System is that if, say, a distribution board load is not balanced (as above Loads A–D) between all of the phases and is, say, not 'resistive', the neutral current can rise significantly and we would end up with a difference between the earth block on the TNC-S system and True Earth. I have marked the difference in Earth Potentials as '**X**'.

This could be a problem in many different cases, but just let me give you one scenario: We plug a metal-clad 230V extension lead into a TNC-S supply, let us say a control building, and run it out onto a plant into a Hazardous, Zoned Area. We now have a situation where the metal of the structure, earthed to Terre, is different in potential to the metal of the temporary extension lead. Touch them together and **SPARKS!**

Also if the neutral became detached the metal case of the socket or whatever would become potentially lethal.

4 – There is another problem with the TNC-S System, hinted at above, and that is if you run a temporary cable out to a marquee or caravan and the neutral became detached then all metalwork could be at a lethal potential. This could also be a hazard for temporary extension leads.

Sometimes people call this a **P**rotected **M**ultiple **E**arthing System (**PME**) which is **NOT** strictly true as with **PME**, the **PEN** Conductor is tied down to earth at multiple points along its route towards the transformer.

This seems to be common for domestic use along with the TT system. If you look at the expected earth impedance here you may be looking in the region 0.4Ω. Earth currents would be fairly high but hopefully this would cause protection systems such as circuit breakers or fuses to operate faster.

The **TNC** system (without the 'S') is different from the TNC-S Earthing System. This system utilises the SWA/Sheath of the cable as the **PEN** conductor and is an unusual system that must **NOT** be used in Hazardous Areas. We could call it an Earth Return. There is a safety problem with this system, as mentioned previously, in that it becomes dangerous if there is a broken neutral.

1) Can you see why a TNC-S Earthing System is not for Hazardous Areas?

2) Do you know of any TNC-S Earthing Systems?

Earthing System TNC:

The pure TNC System consists of an Earth from the Star Point of the Distribution Transformer **BUT NO NEUTRAL** as such. At first glance it looks like a TN-S Earthing System, but I can assure you **IT IS NOT!** THIS SYSTEM IS NOT SUITABLE FOR A HAZARDOUS AREA.

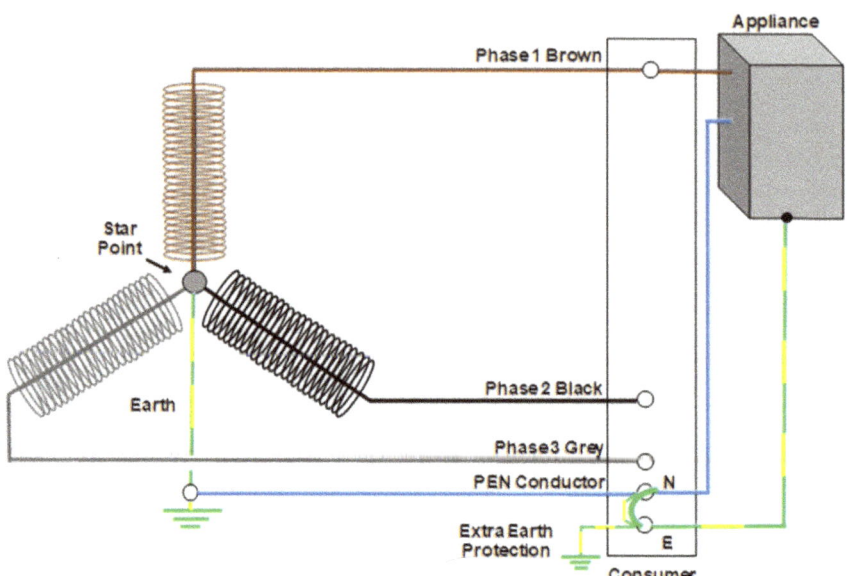

If we look at the diagram above we can see that the star point of the distribution transformer is earthed so this part of the system is **TN** because it is 'neutral' if you can call it that; if connected together with the Earth then this is the **C – Combined**. The electricity supplier will probably use the SWA of the consumer cable as the Earth & Neutral Return. Similar to the TNC-S Earthing System.

1 – If we now have a break (open circuit) in the Neutral between the electricity provider circuit breaker and the consumer as above, which if they used the SWA of the consumer feeder cable would be unlikely, we have now got a **VERY DANGEROUS** situation.

When switched on our appliance would not work because of the broken Neutral/Earth **PEN** Conductor, so the electrons cannot get back to the star point of the transformer. Because the earth is linked to the neutral after the break the case of the appliance would be acting as the neutral waiting for a path back.

If someone (our stick person) was to touch the metal case while standing on earth, as in the TNC-S Earthing System, then this person would become the neutral and provide the path back through them and the earth on which they are standing, along the ground and up the star point earth to the star point of the transformer and invariably would receive an electric shock which could be fatal.

2 – It is also dangerous in another way. Core Balance units of course would not work and protect anyone as there is no fault to earth and the cores would still be balanced with the fault path.

There are numerous faults with this TNC Earthing System which make it unsuitable for Hazardous Areas. At first glance it may look like a suitable system and under certain circumstances I am sure that it is.

Earthing System TN-S:

On the previous pages we have had a look at the TNC/TNC-S Earthing System and decided that this system IS NOT SUITABLE FOR A HAZARDOUS AREA.

Now let us look in detail at the TN-S Earthing System as this system IS SUITABLE FOR A HAZARDOUS AREA and is one of the most common.

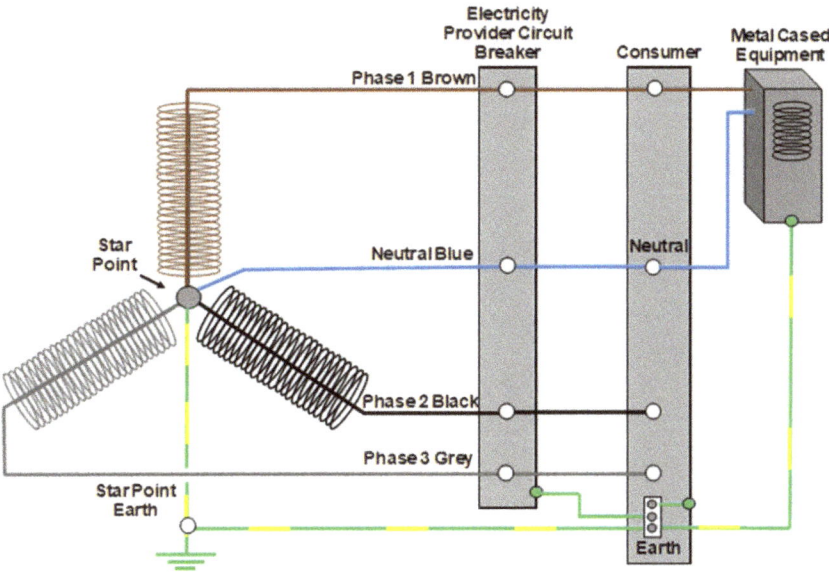

The **TN-S** Earthing System is one of the older systems, but safer for our Hazardous Area. This system does not rely on the earth block being supplied off the system neutral so the block should be at the same potential as the star point earth at the distribution transformer.

The Earthing is supplied from the electricity supplier to the consumer via a separate core/**SWA of the feeder cable**. What do the letters actually mean? **T** = Terre; **N** = Neutral (TN = Star Point Earthed); **S** = Separate Earth Wire.

On a **TN-S Earthing System** we have the **TN** part which is back at the Star Point Earth and the '**S**' because our earth wire at the consumer is Separate from the neutral.

In this system the resistance/impedance would be much lower so a larger earth fault current may flow momentarily but the protection would hopefully act quite quickly.

If you look at the expected earth impedance here you may be looking in the region of **0.8Ω**. The TN-S Earthing System would, if correct, supply enough **Prospective Fault Current** to trip the protection system. In a domestic situation care should be taken when choosing the B, C or D type Circuit Breakers to ensure that the Fault Current value, worked out from the Earth Loop Impedance, is enough. (See later section on Earth Loop Impedance.)

1) Do you see the problem with incorrect circuit breakers?

2) Have you got a TN-S Earthing System on your plant?

3) Have you analysed the Earth Loop Impedance readings?

4) Has the prospective earth current been worked out for your plant?

Earthing System TT:

We have had a look at the TNC-S Earthing System and decided that this system **IS NOT** SUITABLE FOR A HAZARDOUS AREA. We decided that the **TN-S Earthing** System **IS** SUITABLE FOR A HAZARDOUS AREA.

Now let us look in detail at the **TT Earthing** System as this system **IS ALSO** SUITABLE FOR A HAZARDOUS AREA.

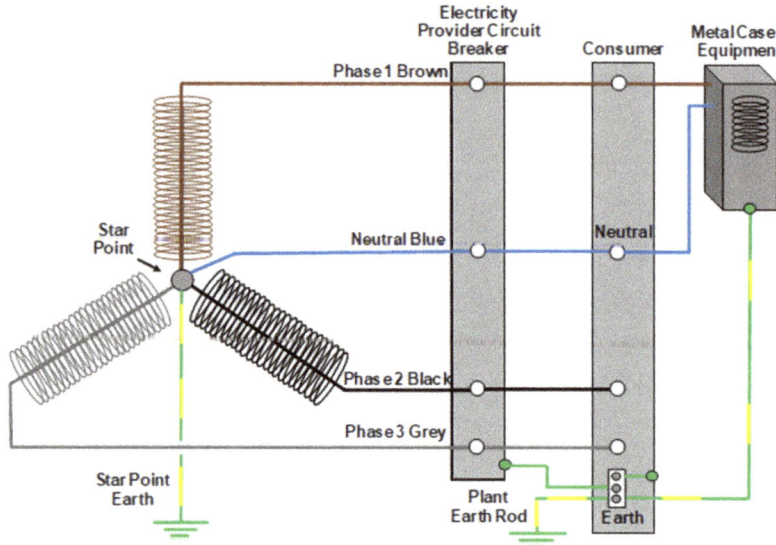

The TT Earthing System is suitable for hazardous areas as the consumer earth is via earth rods in the ground. It is known as a **Terre Terre (TT)** System and, provided the consumer earth rods are tested correctly and regularly, the transformer star point earth will be at the same potential as the rods at the consumer. Any number of rods can be used at the Consumer end.

Looking at the diagram above, any fault current to earth, say the metal case of the equipment, will flow down the earth wire, through the consumer rod, through the ground to the star point earth and hopefully operate the protection which may be a circuit breaker or fuse. If the resistance/impedance of the fault loop i.e. transformer winding, phase wire, earth wire and rod to rod came to around 15–20Ω then going by Ohm's Law the current flowing here would be around 11–15 Amps which is not too excessive. So the ground resistance rod to rod would come into play here.

There are drawbacks even to the TT Earthing System that may have to be supplemented by ELU/RCD. Our problem with the TT Earthing System (if you can call it a problem) is that the **Prospective Fault Current**, which is the maximum current required to trip the protection, would be quite low when testing Earth Loop Impedance Ze or Zs and would not be enough to trip the circuit breaker so this may have to be supplemented by the use of Earth Leakage Units (ELUs) of Residual Current Devices (RCD) so that there is a milliamp trip for protection. (See later sections on Earth Loop Impedance.)

1) Do you see the problem with incorrect circuit breakers?

2) Have you got a TT Earthing System on your plant?

3) Have you analysed the Earth Loop Impedance readings?

4) Has the prospective earth current been worked out for your plant?

Earthing System IT:

We have had a look at the **TNC-S Earthing System** and decided that this system IS NOT SUITABLE FOR A HAZARDOUS AREA. Now let us look in detail at the IT Earthing System: IS THIS SYSTEM SUITABLE FOR HAZARDOUS AREAS? I will let you decide after reading the description below. The question was asked at the beginning: 'Why don't we take the star point earth off?' Well in this system in some cases we have done that, in others we have inserted a high impedance.

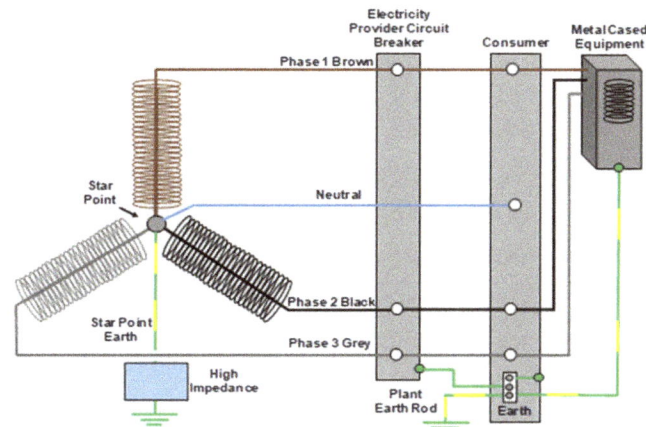

Because there is a high impedance, or no connection to earth at all, in the transformer star point earth the fault current at the consumer is cut down enormously. There are fewer chances of fires or explosions because of the low or non-existent earth current. Equipment also does not suffer the damage of high current. This system mainly uses a three-phase feed to equipment, but non-earthed neutral can be used.

I have not mentioned 'I' as in **IT** Earthing System until now and here the 'I' stands for **Isolated** and 'T' always stands for **Terre**. It is used in situations like mines, operating theatres, ships, platforms etc. where a constant power supply with no power cuts is vital. The chance of what is described below actually happening is quite slim.

The question must be asked: 'by removing the star point earth on the distribution transformer on this IT system, does it rely on the fact that the feeder cable neutral **NEVER** gets a fault to earth? If this were to happen would this system suddenly revert back to a TN-S or worse a TNC-S when the neutral is earthed, i.e. spike through the neutral?'

One plus about this system is that the high impedance in the star point earth means that if we put our stick person into the diagram they would not get a shock to earth of any magnitude because there is no direct reference to earth there. One minus is that two simultaneous faults could cause a current problem making the protection system trip. This could be dangerous depending on whether the faults were to earth.

The following **current limiting** systems can be in the distribution transformer star point earth:

1 – A High Impedance: Allowing tripping only with a second fault (IT as at top.)

2 – A Petersen Coil: A tapped iron core coil to limit capacitance earth fault current

3 – A Reactance Earth: A reactor is placed between the transformer & earth

4 – A Resistance Earth: A resistor is placed between the transformer & earth

RCDs are not applicable on this system because there is no direct path back to the Distribution Star Point.

1) Have you got an IT Earthing System on your plant?

2) If 'YES' have you got ELUs/RCDs backing up the system?

Earthing Methods:
Rods/Plates/Lattice/Electrolyte/Electrodes/Coils:

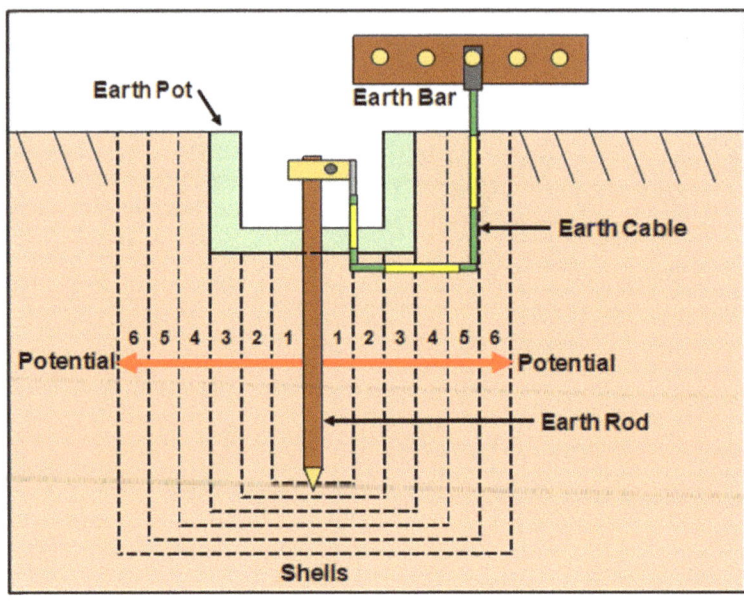

Now that we have discussed earthing 'Systems', which of the above earthing 'Methods' is most suitable for each particular situation? Remember what we are looking for is surface area so plates and lattice might provide that, but in certain situations just a rod may be ideal.

There are many examples of different types of earthing methods and some advantages of one against another. I will let you read the following sections and then we will see the pros and cons around having the different methods installed. Do we use electrodes, plates, coils or lattice?

When building a substation or switch room it might be, for instance, that due to the ground resistance tests we build an enormous lattice and site the building on top of it, allowing several points to surface in an earth pit for testing.

As we will discuss, many earth electrodes i.e. rods, are not made of copper as many people think. They are actually steel rods which are copper coated as we will discuss later in this section.

1) What Methods have you discovered on your factory?

2) How often do you test your Earthing Electrodes?

3) Are all of your Earth Pits identified with colours/numbers?

4) Are all Earth Bars Identified with numbers?

5) Are all Earth Cables identified at the Earth Bars as to which electrode?

6) Have 'Step Voltages' been identified?

Standard Earth Rods:

There are several forms that earth protection can take. Providing earth protection with whatever method is all about surface area in the ground. Let us start with **Earth Rods**, these are a quick way of achieving just one earth protection. They can be hammered quite easily into the ground and are around 120cm (4ft) long. Most Earth Rods of course are **made of steel and only copper coated** and not solid copper as some people think. If the rod was made of solid copper, being a soft metal, you would have problems hitting them into the ground, especially if you had to go a few rods deep, and of course would be very expensive. Rods can also be galvanised steel or stainless steel. Sometimes a **'Pipe'** electrode can be used instead of a rod. This pipe electrode is quite common and is around 30-40mm diameter and around 3m long.

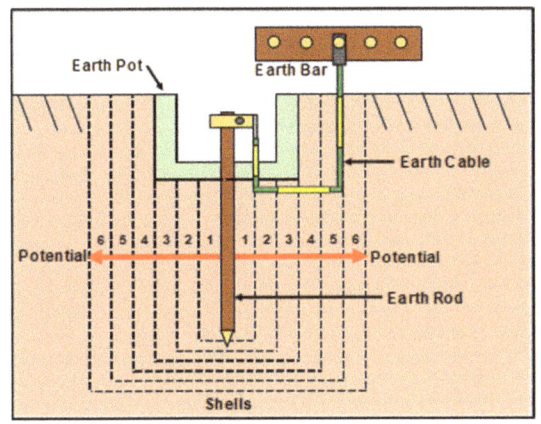

First of all in the diagram, left, is our earth rod driven into the ground (**Earth Pit**) with the **Earth Pot** to protect it at the top. The Earth Pot would be made of concrete on a factory complex so that vehicles did not damage it and have a slightly tapered lid to cover the top of the rod and seal the pot. The earth cable would be connected to the rod and a plant earth bar and there would be several of these around the plant all connected together.

The 6 dotted zones, as in my diagram, are called **'shells'** which are different sections of ground which form the path of potential, marked in red, away from the rod. They would of course go all the way round the rod in a radius of several feet. It is possible for these 'shells' on the ground surface to differ in potential quite substantially in the event of an earth fault. How big the 'shell' radiuses are depends upon ground resistance and voltage. Step voltage would have to be looked at here.

If you drove another rod into the ground within the 'shell' radius then this would be classed as the same rod as the first one when the two are joined together. Where I worked at BP we had three rods or pipes fairly close together, connected them in a delta wiring formation and called them an earth rod 'cluster', and this was done purely for surface area to get a reading of 1Ω. Other separate system rods should be at least around 2–2.5 metres away from the first to get outside of the 'shells'. On very high voltage, say 33,000V, the 'shells' may be hundreds of volts potential different on a large earth fault and would be enough to kill wildlife should they be standing in two different shells on the ground surface at the time of the fault. Remember we are looking at **surface area** rather than depth, although the deeper the rod, the more surface area.

The deeper the rod the more cross-sectional area of metal is in the ground and the more fault current dissipation, but also the closer to the water table the less resistance there will be. It is possible to lay the earth electrode horizontally, but it may have to be a pure copper rod at least 6.1 metres long and 13mm diameter and laid at a minimum depth of around 1–2 metres.

1) How many Earth Rods do you have on your plant?

2) Are any of the Earth Rods vertical or horizontal?

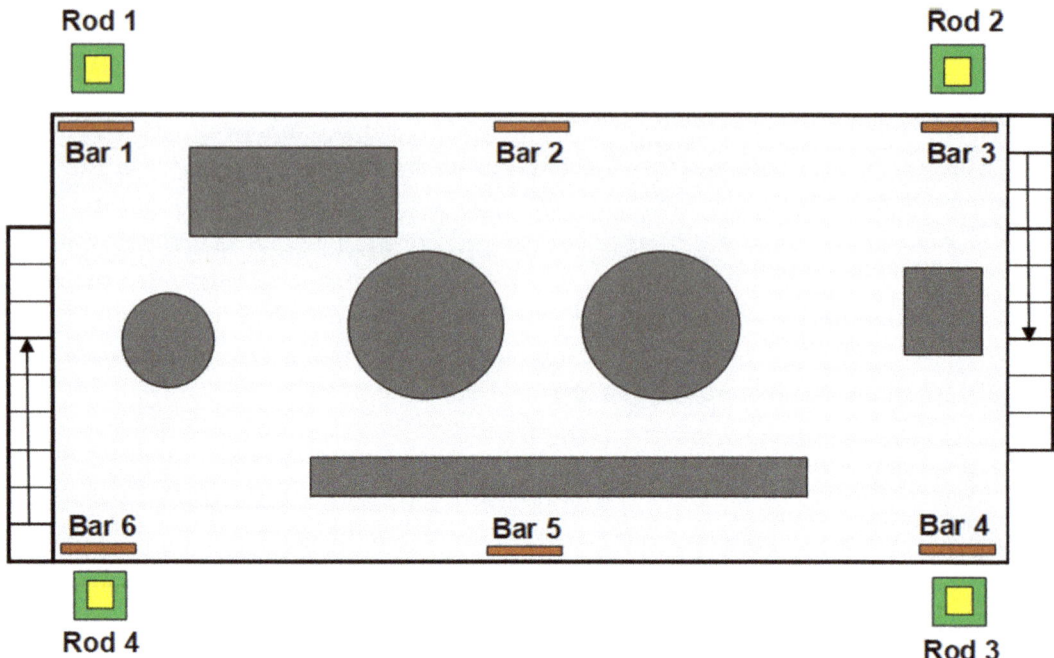

The above diagram shows how the Earth Bars and Earth Rods may be laid out on the ground floor of a chemical complex plant. As you can see there are six Earth Bars, one on each corner and two in the middle. These bars would be connected together in a large ring with a green and yellow sheathed cable, something around **70mm**. These would also be connected to Earth Bars on the upper floors. Just remember what you are trying to achieve: a path back to the star point of the distribution transformer in the event of a fault to earth from an electrical source.

Also above you will see an Earth Rod on each corner of the plant. We actually painted the Earth Pots on the concrete green and yellow as above to make them stand out. The Earth Rods connect to the Earth Bars on the four corners of the plant and we would be looking at an Earth Rod resistance of **5–10Ω**.

All of the vessels and columns, represented by dark grey shapes, were then connected to the Earth Bars, usually with one earth connection per vessel, with a tank that may go up to two or three. The girders that actually make up the structure of the plant would also have an earth connection to these bars. By doing this we can ensure that all metalwork on the plant is at the same potential. **(Equipotential Bonding.)**

Also, on a structure this size with a path to the ground for the lightning potential, just remember that here we are just trying to balance the ground potential with the strike, nothing to do with star points of distribution transformers. Lightning conductors positioned on the highest structure on the plant might go to different Earth Rods. The reading for these Earth Rods would be in the region of around **10Ω**. The resistance for lightning protection does not have to be great so long as there is a resistance there.

Most Chemical Complexes would have the TT Earthing System for the site and possibly a TN-S Earthing System for offices and stores.

1) Is your plant Earthing System similar to the one above?

2) What is the average acceptable Earth rod resistance?

Earthing Plates:

Earthing Plates are thin plates which can be made out of a coated ferrous material or pure copper and provide a good surface area which is in direct contact with the ground. If we were to take a building like a substation, large earth plates can be put into the ground on the corners around 3 metres down. Being buried in the ground the plates are obviously open to corrosion, especially if they are of the ferrous metal type.

Now we join the earth plates together sideways, lengthways, and corner to corner and form an 'Earth Mat' so we can get even more surface area.

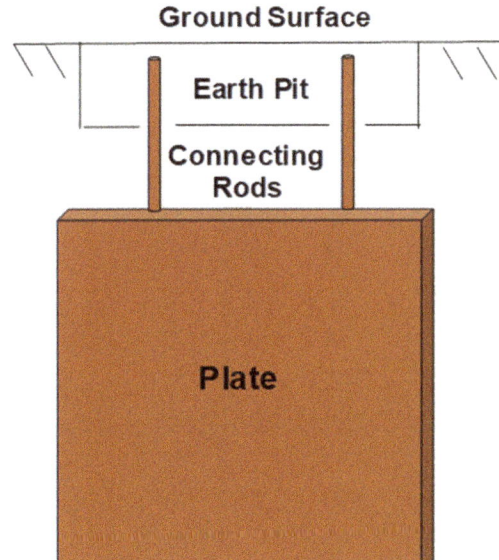

The plates are best vertical, as in the diagram left, to capture more ground layers, and **from** 60cm (2ft) square up to what you like, but remember if these are made of copper and not ferrous metal that could be very expensive. Ferrous materials would require to be thicker than copper, around 7mm, where copper need only be around 3mm. Obviously the number of plates required and the size must be worked out carefully.

One problem which probably stands out above the rest is corrosion, and care must be taken where any rods or earth leads connect to the plate, to ensure that the joint does not corrode off. Depth: **Vertical** around **3 Metres** from ground level. Certain 'Backfill' may cut down corrosion and help conductivity.

Plates might need to be considered where the terrain might be rocky or hard and driving rods into the ground might be difficult.

The plates can be flat in the ground with 1/2 earth leads connected to them. As shown in the diagram to the right the plates are around 60cm x 60cm. The holes which house the plate if laid flat are roughly 75cm deep. Powdered charcoal and lime mix are used to cover the earth plate vertical or horizontal to a depth of around 30cm. In some areas with a different water table the depth of hole can increase by as much as 3m.

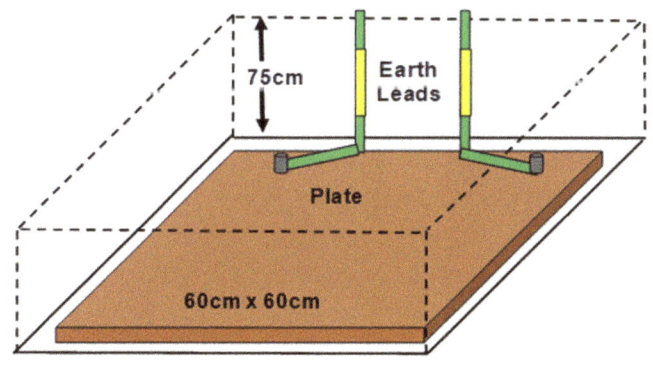

The filling medium, mentioned later, helps both with conductivity and moisture retention around the Earth Plate itself. Just be careful which medium is used as sometimes this increases the corrosion which is the opposite of what you want. There are many sites on the internet as well as manufacturers of the plates for advice. Filling mediums are usually materials such as clay that retains moisture. These compounds lower the soil resistance around the electrodes, thus helping the electrode make better contact with the surrounding ground, be it soil etc. This might be a consideration on high voltage installations. **For backfill compound, electrode size and depth, manufacturers' advice is essential.**

Earth Lattice:

An Earth Lattice may be used where knocking earth rods into the ground may be extremely difficult. Remember, what we are trying to achieve here is surface area. Several of these lattices are laid flat in the ground and connected together and may not be quite as expensive as solid copper earth plates. Which system is used depends upon the initial soil resistance test using the 4 terminals on the tester.

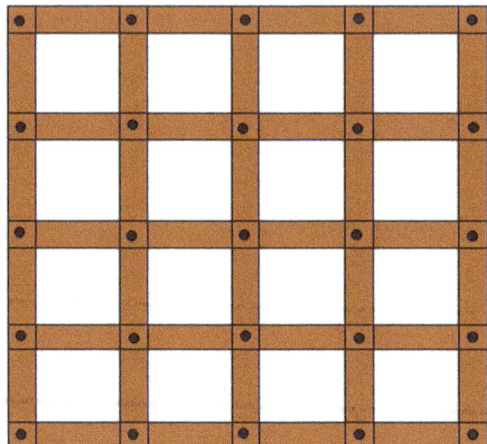

The lattice can cope with high fault current and has the surface area. Another name for this type of protection is **'Earth Mat'**. It might be that many lattices are laid flat and joined together to make a huge mat for the switch room or substation to sit on top of it.

The lattice can be made up of flat earth strips, as in the diagram above, or round Earth Rods, which might be a bit more expensive. On a high voltage substation there may be several of these lattices all joined together. This, of course, would have a huge surface area and be less expensive than copper earth plates. Earth rods may be driven into the ground and electrically connected to the lattice in several places. This would bring the earth connection up to ground level and enable substation/switch-room earth bars to be connected to them. Again, with a very high voltage substation, we would have to ensure that the Earthing 'Shells' (different parts of ground around the earth conductor) did not pose a danger to personnel in the event of a large earth fault. The size of the lattice would typically be from 60–90cm and give a surface area of around 0.5 square metres each.

If there are poor initial readings of ground resistivity when the preliminary testing is done there are certain 'Soil Conditioning Agents' that can be applied to assist the resistivity, such as Marconite & Bentonite. Marconite is more permanent and can be mixed with the cement surrounding the lattice. Also more rods can be inserted and connected to the lattice. Filling mediums are usually materials such as clay that retains moisture. These compounds lower the soil resistance around the electrodes, thus helping the electrode make better contact with the surrounding ground, be it soil etc., and cut down corrosion. This might be considered for instance on high voltage installations. There are plenty of compounds on the market and **manufacturers' advice is essential**.

1) Do you have Earth Lattices & Earth Mats on your plant?

2) How often are these installations tested?

Electrolytic Earth Electrode:

Earth **'Electrolytic Electrodes'** are an unusual approach to earthing. Instead of using solid metal earth rods, a conduit-like tube is used which is full of a special concoction of salts and materials that absorb water from the air using an upper cap with holes in to allow moisture to enter and mix with the chemicals.

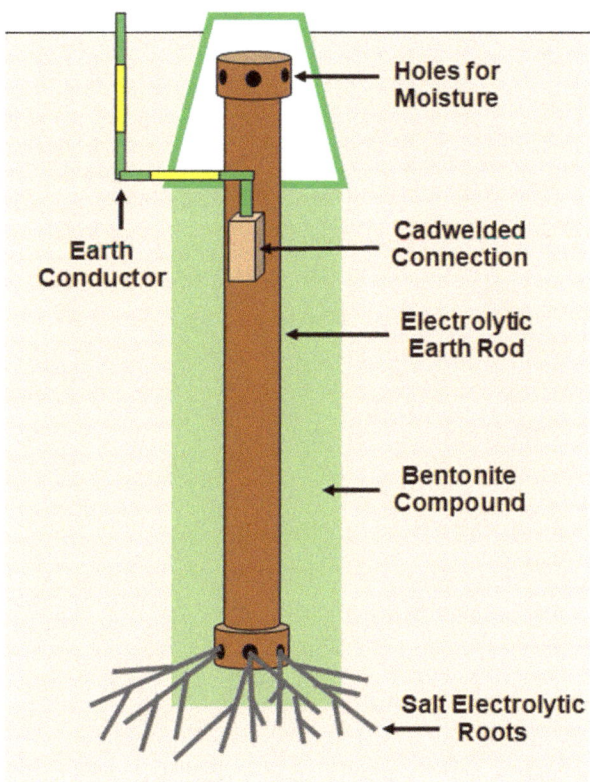

As shown above, what we have here is a tube full of electrolyte so there would be no driving it into the ground with a hammer. It would be put into a hole, possibly made with an augur. The tube has the earth cable fastened onto the side using a **'Cad-weld'** which is a special mixture of flammable powder like gunpowder mixed with a substance like brazing in a mould around the joint. This compound is lit and it welds itself to whatever. The tube has a cap on the top which would be visible inside the earthing pot. This cap has holes in to allow moisture in and keep the electrolyte fluid.

There is also a cap on the bottom with holes in so that the salt from the electrolyte can soak out making the tube look like it has roots. These give the electrolyte contact with the surrounding medium. The whole tube is then surrounded in the ground by a medium material such as bentonite clay to give a highly conductive path to the ground as well as protecting the tube from substances in the soil that may cause it to corrode. The tubes can be vertical, as in the diagram above, or flat, in which case a trench would have to be dug to house them. They are used where there are poor ground/soil conditions. Several tubes can be joined together in the ground and usually each tube would be one metre in length. As with the earth rod explained earlier, this earth electrode will also have 'shells' which, as previously mentioned, are different areas of ground potential going away from the electrode. **Manufacturers' advice must be obtained for installation & testing.**

1) Have you got any Electrolyte Earth electrodes on your plant?

Earth Electrodes in Concrete:

It is possible these days to obtain earth electrodes encased in concrete. There are certain rules on how thick the concrete has to be and the size and length of the electrode.

When we talk about electrodes, plates, lattices etc. what we are looking for is surface area. When putting earth rods into the ground it is not how deep we go, but how much surface area of copper we have in the ground, so lattices and plates will usually come out top. Now if we look at the method above and we use a conductor which is around 15mm in diameter and 'X' number of metres long, usually a minimum of around 7 metres (around 23 feet), then this will give a good surface area.

One problem with **older type** concrete encased electrodes is that when the concrete is in the ground it absorbs water. Now, remember why we have this electrode in the first place, for safety to allow current to flow back to the distribution transformer in the event of a fault. If the fault was a large one and quite a lot of current was to flow through the conductor, much heat would be produced as well as stresses and the water in the concrete could then turn to steam, thus expanding and exploding the concrete, maybe damaging the conductor as well. If the steam explosion happened and the concrete was severely damaged but the conductor was still intact, then all would still work except now the conductor with no casing would be just like any other conductor we hammer or lay into the ground but with not so much corrosion protection.

On more modern concrete encased earth electrodes, carbon or earth conducting backfill can be added to the concrete mix making it much more conductive to the surrounding ground. Adding carbon or something like charcoal and making the resistance less would substantially cut down the risk of steam damage being created from a large earth fault due to water impregnating the concrete.

One mass of earth which could, of course be explored is any steel reinforcement in concrete in, say, the base of a building. This would supplement an electrical earthing system to a huge degree if it could be accessed after the building is built. This would form a huge grid and could be connected to strategically placed earth rods or plates.

Remember what we have been saying about resistance values in previous electrical earthing in that we are looking for a value of around **5Ω** or better.

1) Have you heard of Concrete Encased Earth Electrodes?

2) If 'YES' do you have Concrete Encased Earth Electrodes?

3) Does your Company connect steel re-enforcement to their Earthing System?

4) What Earthing Test values are achieved?

Earth Coil Electrodes:

Earth coils at the moment are not the most popular earth electrode, but the theory of them is excellent. Because of the coil the amount of surface area which can be obtained here is huge in a relatively smallish actual area.

The coil may be much thinner than my drawing with many more turns and can be laid flat if required. Of course a hole would be dug for positioning of the coil either vertically or horizontally, depending upon manufacturer design.

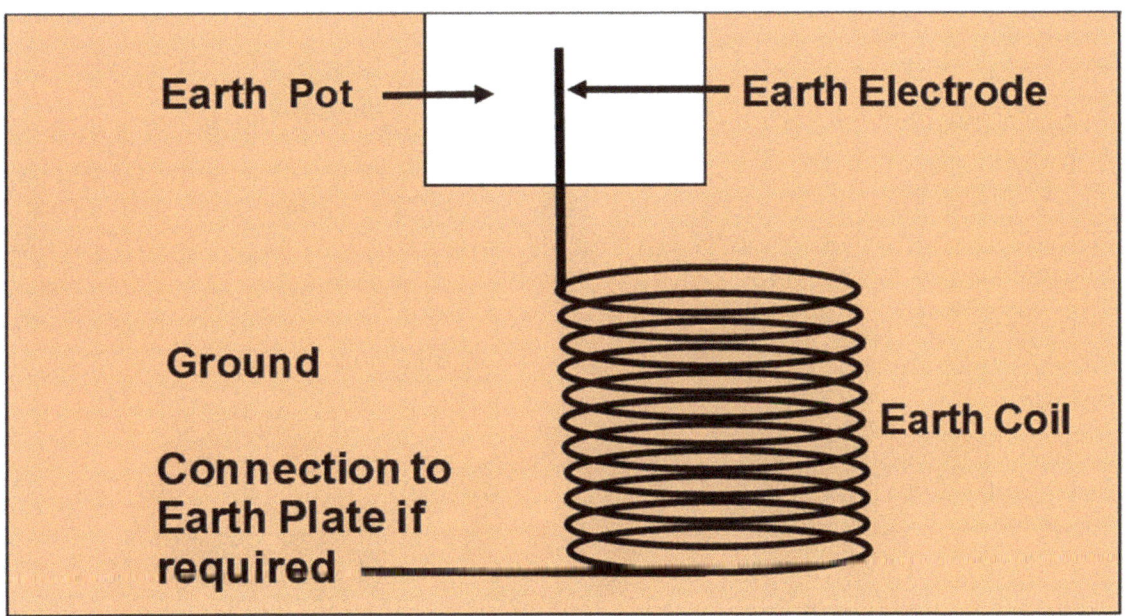

We can take a wire that is around a minimum of 5 metres long and make it into a coil as we have done above. The coil would be around 5cm wide (approx. 2"). The number of turns would be determined by the length of wire and the diameter of the coil. The coil would be vertical or flat and, if the latter, would be at a depth of around 1–2 metres. The manufacturers would give guidance on the latter dimensions.

The coil can be made of galvanised iron, mild steel or copper and again, as with many of the electrodes we have mentioned, the bottom of the coil is surrounded with charcoal or special Earthing Backfill.

This would be an ideal situation for a 'Cadweld' jointing system. The coil can protect HV systems as well as MV & LV.

Filling mediums are usually materials such as clay that retains moisture. These compounds lower the soil resistance around the electrodes, thus helping the electrode make better contact with the surrounding ground be it soil etc. This might be considered for instance on high voltage installations. There are plenty of compounds on the market. **Manufacturers' advice is essential.**

1) Have you heard of Coil Earthing Electrodes?

2) If 'YES' does your plant have any?

Earth Rods and Accessories:

Many people think that Earth Rods **(No.2 below)** are actually made out of copper. This of course is not so. Pure copper would be very impractical because not only would it be too soft to hammer into the ground, but it would be very expensive. They are in fact copper-covered steel.

The copper coated steel earth rod is threaded at one end to allow the brass coupler **(No.3 above)** to be screwed on to enable another earth rod to be attached should a greater depth be required. The intermediate rod would of course be screwed at both ends. The coupler also allows a bolt **(No.4 above)** to be screwed into the end so that neither the rod nor the coupler are damaged during hammering into the ground. (Sometimes this is a rounded bolt and is called a 'driving head' and if in good condition at the end can be left in to protect the inner thread.) A 'point' **(No.1 above)** can be screwed into the end of the earth rod that is going into the ground to make it easier to drive the rod into the ground. (Sometimes this is called a 'driving tip'.)

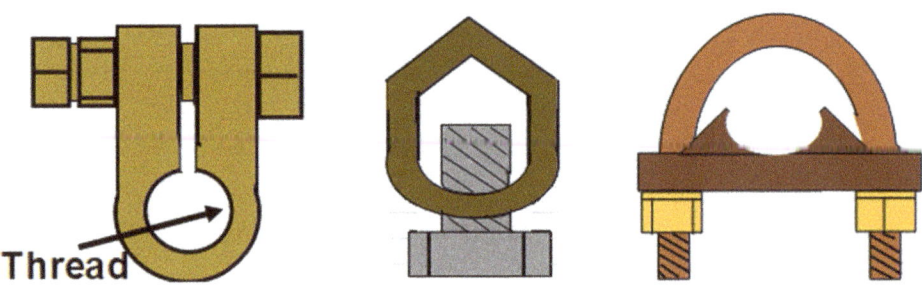

Above are three different style earth wire clamps (there are more!). The clamp on the left has a screw thread on the inside so that it can be screwed onto the top of the earth rod to allow the earth cable to be connected to the rod. This seems the most convenient. The clamp in the middle passes over the top of the earth rod and clamps the cable using the bolt. The clamp on the right is similar to what we used to call bulldog clips used on catenary wire.

These clips are used to enable easy disconnection of the earth cable from the rod to enable conventional earth rod testing to be completed. This would be the last job when the earth pit and earth cable have been installed.

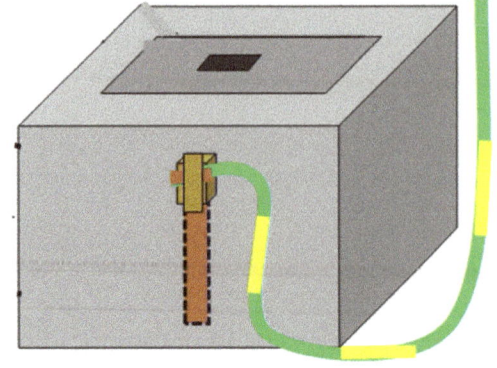

Finally there is the **"Earth Pot'**. This can take the form of a small black plastic box which fits on the end of the earth rod which is not all that convenient as the rod is below ground level.

Another type is what looks like an inverted plastic litter bin with a lid which might be ok for domestic or light duty. For heavy industrial use where vehicles may be going over the top of it then a very sturdy concrete type is available. See later section on Earth Pits.

Joining Earth Electrodes:

Let us just say that we have two copper earth tapes, rods, lattices, plates or wires etc. that require joining or copper equipment joining onto them, what are the different ways that we could do this?

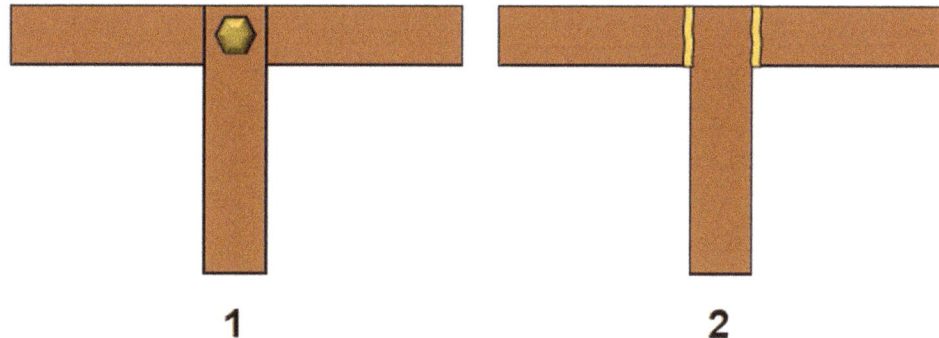

Let us stick with earth tapes for ease of this explanation. Firstly we could, as in **diagram 1** above, just drill them and put a nut and bolt through. I suppose the nut and bolt could come loose and form a high resistance joint.

The more that are done like this the more chances that one or more could loosen. We could, as in **diagram 2**, braze or solder them together, but if there are numerous to do then this could take time and if we are in the middle of our Hazardous Area, gas free certificates for the solder or brazing device would be needed. With each of these methods we are relying on human action.

There is another option that is extremely efficient and that is completing what is known as a 'Cadweld'.

Now this would be extremely difficult in the middle of our Hazardous Area as it would be like a firework and would involve gas free certificates and a very stringent Risk Assessment, Method Statements and Permit System. Let me try and explain how it works below.

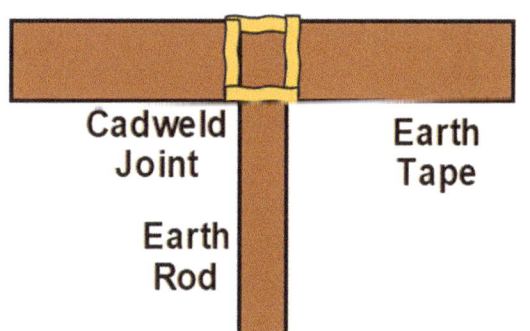

The 'Cadweld' mould and accessories are put around whatever shape the joint is i.e. wire to wire, wire to rod, tape to tape etc. The mould is filled with a gunpowder type powder. The weld material is in the accessories that are inserted first. It is very important at this stage that safety equipment is put on by the user i.e. safety glasses, gloves etc. The powder is ignited by an igniter (spark gun) and there is a huge flash.

When the mould is removed the earthing tapes, rods etc. are welded together and would require to be hacksawed apart. OK for substations and switch rooms without too much documentation, but although this is a very efficient joining method, Zoned Areas do not like fireworks in the middle of them so using this option with the plant running may be limited.

Many people think that the 'Cadweld' is a new technique of joining coppered earthing components but it is not. It was demonstrated to me during the building of our Monsanto Plant in the 1980s.

Earth Pits:

An **Earth Pit** is simply a hole in the ground which facilitates earth electrode. At the top of the earth pit is the end of a rod where the earth cable connects to the conductor. The hole housing the electrode can then be filled with earthing 'backfill' material to help the conductivity. On the top of the earth pit, as mentioned, is an earth pot which protects the top of the electrode and can be made of plastic or concrete. The worst part about earth testing is getting the wedge-shaped concrete lid out of the earth pot if it has not been removed for years.

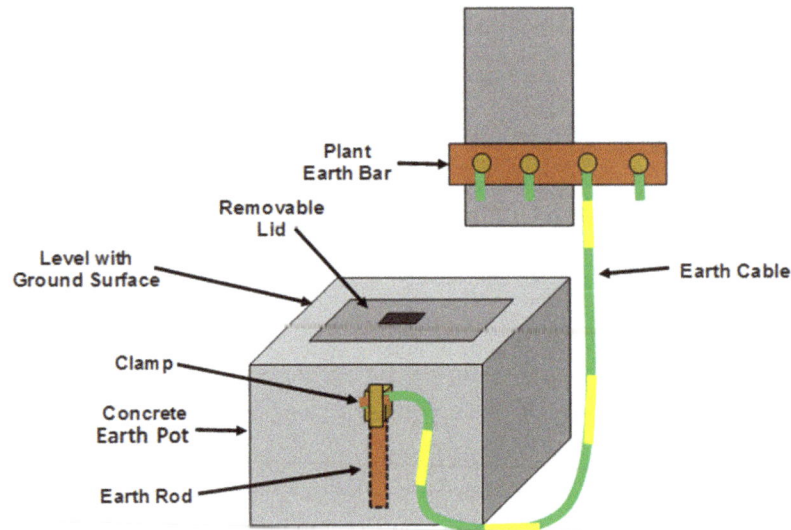

If there is going to be heavy plant such as trucks, cranes etc. riding over it then usually the earth pot will be made of concrete with a wedge-shaped cover pushed in the top.

After a few years, the lids of these pots are extremely difficult to remove to get at the earth rod for testing and we ended up renewing every single one on the plant which meant guys digging the original ones out with a pneumatic drill and concreting the new ones in.

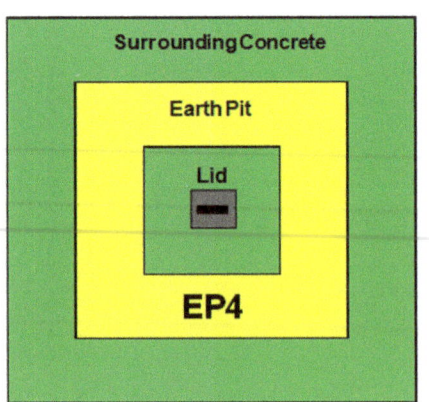

One thing that we did do on our Chemical Plant was to paint the earth pots/pits with green and yellow **non-slip paint.** There were around 20 earth pots altogether on each plant being joined to various earth bars around the structure by a 70mm earth cable.

Each earth pot/pit had a unique number, EP1 to whatever, and drawings were done to show the position of each pit. It is advisable to paint the sides of the lid with a grease type substance so that they can be more easily removed in the future.

Once the earth pots were sorted out it was just a case of lifting the lid to ensure that all was OK at the top of the earth rod and testing the rod resistance with the clamp-meter which cut the testing time down enormously.

1) Can you remove the covers on your Earth Pots?

2) Do you identify your Earth Pots?

Earthing Conductors:

So what can we use as an earthing/protective conductor according to the IEE Regulations? Below are the examples that you might find in your hazardous area:

1 – A single core cable. This of course is used to connect an earth rod to a plant earth bar. The single core cable insulation should be green and yellow. **(Usually 4mm minimum – 70mm.)**

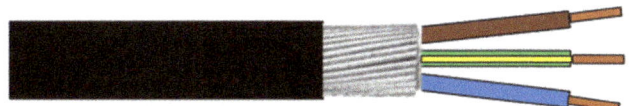

2 – A Conductor in a cable. Some SWA & braided cables have earth conductors and do not use just the SWA for earth although of course the SWA must be earthed. Universal or dedicated 2 seal glands with SWA trap must be used. These glands must be made off complete with accessories such as 'star' washers, IP washers where required, locknuts etc., by competent technicians. Glands can be obtained to certified protections such as Universal, Exd Flamepoof, Exe Increased Safety, Ext Dust Protection and ExnR Restricted Breathing.

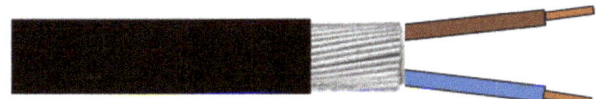

3 – The cable above is a 2 core and the SWA or braid is being used as sole earth. Universal or dedicated 2 seal glands with SWA trap must be used. These glands must be made off complete with accessories such as 'star' washers, IP washers where required, locknuts etc., by competent technicians. Glands can be obtained to certified protections such as Universal, Exd Flameproof, Exe Increased Safety, Ext Dust Protection and ExnR Restricted Breathing.

There have been instances in very damp conditions where the gland thread has rotted off or the SWA has become detached from the gland, thus leaving the metalwork unearthed. If equipment requires an earth, in my opinion, it is always better to use a 3 core cable with an earth core. It may be a company policy that the SWA is not used as the sole earth and a third core must be run.

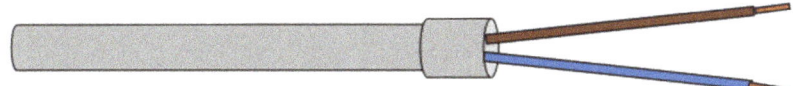

4 – The regulations do allow for a conduit, trunking, cable tray or ladder rack etc. to be used as an earth. Just be careful that the way the metalwork is earthed is sufficient and robust and cannot be easily removed, damaged or broken off. Conduit is not used in the quantities it once was so that problem is not usually present. If the cable tray or ladder rack is going to be used as an earth return, and is passing through a Zoned Area, then the Electrical Engineer may insist on braided bonds across every joint on the run. Conduits can break and hence break the earth continuity and systems may have to be inspected at regular intervals. The weakest part of the conduit, of course, is the thread.

Earthing Electrode Backfill Compounds

Backfill compounds are placed in the earth pit around earth electrodes to make the immediate ground close to the electrode more conductive, usually used when the terrain is not too conductive i.e. granite.

Some compounds have huge water retaining properties for long lengths of time and are backed up by ground water.

There are several backfill compounds on the market which can be used with any method of earth electrode. They are eco-friendly and non-corrosive. I have listed several below but there are more:

Marconite:

Marconite is a popular electrical earthing electrode backfill compound, made up of several carbon compounds, which has a very low conductivity.

It is mixed with a smaller amount of cement and has the strength of concrete when it has set.

We are looking at a low resistivity of around 0.1Ω/metre.

Bentonite:

Bentonite differs from Marconite in the fact that it is a clay consisting of montmorillonite (a group of compounds that form crystals & gel when introduced to water).

There are two types of Bentonite i.e. Sodium Bentonite and Calcium Bentonite. The latter may be called 'Fullers' Earth'.

Slightly higher resistance than the above-mentioned Marconite at around 2Ω/metre.

Bentonite also protects, to a certain extent, the copper/copper-coated electrodes from acids and alkalines in the ground.

Gel:

Gel compounds are rare earth compounds that usually have a very high 'hygroscopic' (water absorbent) property and will hold water for considerably long periods and are highly conductive.

The conductivity of the gel is around 0.1Ω/metre.

3) Have you heard of earth electrode 'backfill' compound?

4) Have you used earth electrode 'backfill' compound on your plant?

5) If the answer above is 'YES' which one?

6) What sort of terrain did you use it on?

Chemical/Gel Earth Rods:

There are several forms that earth protection can take. Providing earth protection with whatever method is all about surface area in the ground. This can take the form of huge earth plates where driving rods into the ground may be impractical, a grid which is criss-crossed from several rods, just one or two earth rods or a mixture of all.

Let us start with **Earth Rods**, these are a quick way of achieving just one earth protection. They can be hammered quite easily into the ground and are around 120cm (4ft) long. The rods of course are made of steel and are only copper-coated and not solid copper as some people think. If the rods were made of solid copper, being a soft metal, you would have problems hitting them into the ground especially if you had to go a few rods deep.

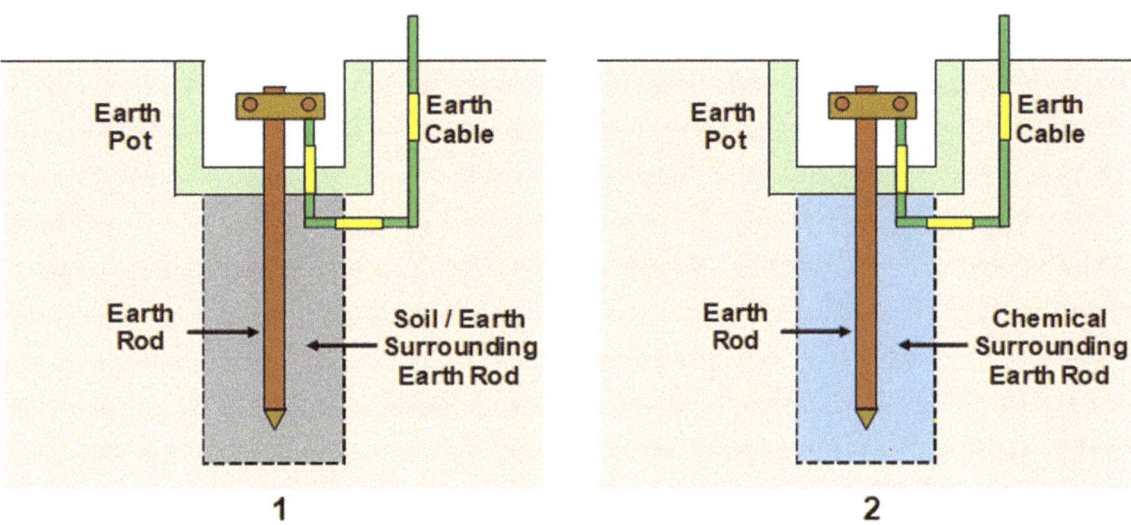

When we talk about chemical earth rods we are not talking about the rod itself, but what is directly surrounding it. In the ground are solvent, salts, acids etc. which tend to start off a corrosion process on the copper earth rod although this would in actual fact, in my opinion, take many years.

Filling mediums are usually materials such as clay that retains moisture. These compounds lower the soil resistance around the electrodes and thus help the electrode make better contact with the surrounding ground be it soil etc. This might be considered for instance on high voltage installations. There are plenty of compounds on the market and **manufacturer's advice is essential.** Sometimes conductive materials like charcoal or some carbon-based substance is placed around the rod to make it more conductive. This set-up would be as in **diagram 1** above.

It is possible to obtain a chemical, sometimes in the form of a '**Gel**', that is put into the ground surrounding the rod, not only making it more conductive, but also cutting down the copper corrosion. This would be as in **diagram 2** above. This process may be used more in Eastern Countries where the terrain is composed of more corrosive materials. Gel is also mentioned in 'Backfill' Compounds.

1) Have you used Gel Backfill Compound?

2) If 'YES' what type of Gel was used?

Soil Resistance Earth Testing:

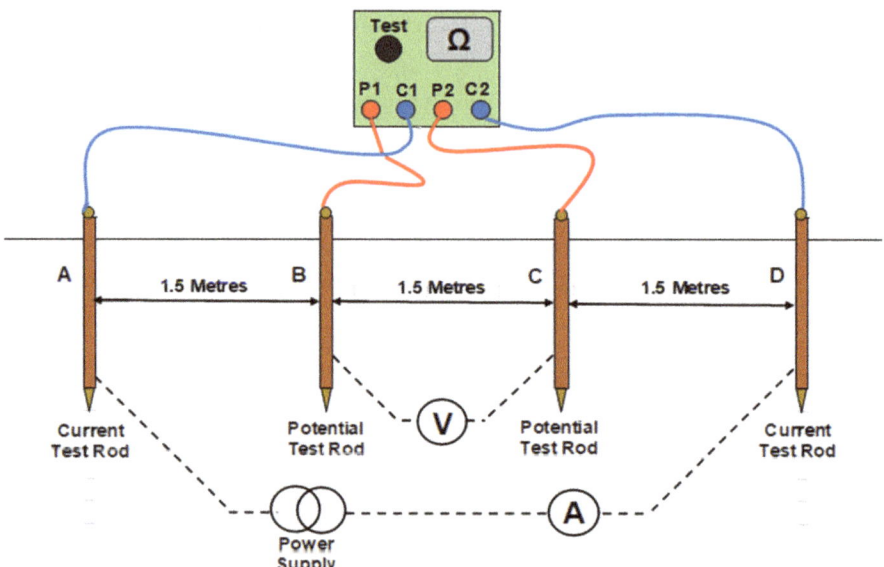

Now that we have discussed what earthing **'System'** might be the most suitable and the ones that are not, and we have decided what earthing **'Method'** to use, the next thing is how do we test the earthing **'Method'** to ensure that there is a good earth?

Firstly in this section let us discuss how we test the ground conductivity as well as the resistance of the earth rod, plate etc. If we were going to build an electrical substation or a switch room in the middle of a plot of land then we may first want to know the conductivity of the ground. Once the conductivity reading is achieved then we can decide on what earthing method we are going to use i.e. rods, plates, lattice etc. It may be that we run a giant earth grid using plates joined together and build the substation on top of it leaving testing points available. **Step voltage** may also have to be considered if the voltage of our substation was quite high, to protect personnel in the event of a high voltage earth fault.

The earth 'Tester' usually has 4 terminals which are all used on a ground conductivity test. Other tests on the earthing methods only use 2/3 terminals, depending upon the company policy for earth testing. These testers started off as **'Bridge Meggers'** using the unknown resistance like a Wheatstone Bridge. These would be wound with a handle. Other testers developed from there really up to the electronic units we have today.

No matter which instrument is used, basically what you have (and I will call them 'Meggers') is two Meggers in one. One 'Megger' looks after the potential loop i.e. **'P'** terminals and the other looks after the current loop **'C'** terminals. I have drawn the loops in the diagram at the top. If the Technician wanted, the 'Potential' and 'Current' terminals could be looped and a reading taken to a known stable earth such as the star point earth of a distribution transformer. This test would be no better or worse than using an earth clamp tester which is on the market today. This would not however be a true test.

1) Have you ever carried out Ground Resistance Testing?

2) Was it done as per the diagram at the top and the write-up?

Soil Resistivity Testing:

This process may be done when deciding to build a substation or switch room on a particular piece of ground, and which particular earthing system is going to be used. The test is very similar to the one used to test earth rods except here all 4 terminals are used on the test set as in the diagram below. There is a stable rod 'A' in the diagram and the other 3 rods are moved equal distances apart.

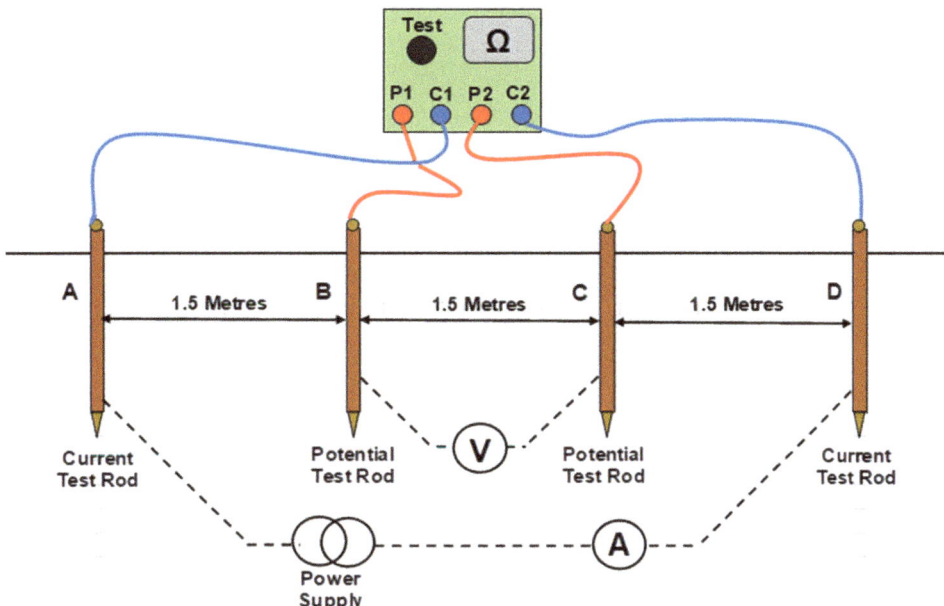

As can be seen from the above diagram, the two outer probes, A and D, are Current Test Rods and go to C1 and C2 on the Earth Test Instrument. I have put an Ammeter and Power Supply in dotted lines to demonstrate what might be going on inside of the Instrument. The other two probes, B and C, are Potential Test Rods and this wiring will go to P1 and P2 on the Earth Test Instrument. I have drawn a Voltmeter with dotted lines to show what might be going on inside of the Instrument here.

The Instrument now has a current and voltage input so we can use Ohm's Law to calculate the resistance in Ohms and put it onto the screen. Looking at the diagram, probes B (**P1**), C (**P2**) and D (**C2**) are adjusted outwards, equal distances apart to the following distances: **3** metres, **6** metres and **9** metres. It is important as mentioned to try and ensure that the probes stay an equal distance apart and try to keep the depth fairly consistent for accurate measurement.

Choosing the particular instrument for the test can also have a bearing on accurate results. The two test meters on offer are High or Low Frequency. Low Frequency Meters seem the most common, especially if you end up with the probes at maximum distance apart, and are less vulnerable to exterior signals which may affect the readings such as power lines etc. What is called **'The Induction Problem'** may affect the High Frequency Models. Manufacturers will always advise and demonstrate the different models.

1) Having completed this test what were your soil resistance readings?

2) What did you complete the test for? i.e. Substation etc.

Ground Resistance Readings:

Below I have shown a chart of approximate ground resistance readings based on the type of soil/rock etc. quantity. These are ballpark figures, just to give you an idea, and alter slightly depending upon which website you look at.

Type of Ground:	Ground Resistance:
Marsh	1 - 30Ω/M
Peat	5 - 100Ω/M
Clay	50 - 200Ω/M
Shale	50 - 300Ω/M
Clay & Sand	50 - 500Ω/M
Chalky Soil	100 - 300Ω/M
Limestone	500 - 5,000Ω/M
Sand	600 - 1,000Ω/M
Rock	1,500 - 3,000Ω/M
Sandstone	1,500 - 10,000Ω/M
Concrete	2,000 - 10,000Ω/M
Granite	10,000 - 50,000Ω/M

The readings are Ohms/Metre (Ω/M) of the above materials. Sometimes the websites will give the readings as Ohms/Centimetre (Ω/cm).

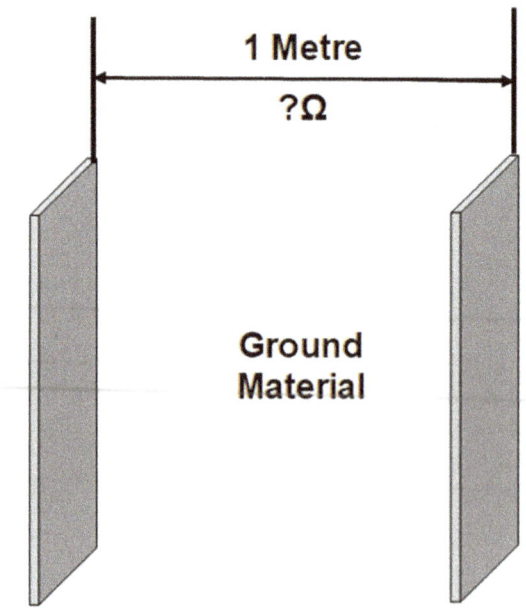

So we look at 1 metre of ground material be that soil, sand, granite etc. and the measurement would be Ω/M.

Testing Earth Electrode Resistance:

This is the conventional method of testing earth electrodes which are either new or existing. If it is an existing electrode then it should be disconnected from the Earth Cable at the clip on top of the rod in the earth pot.

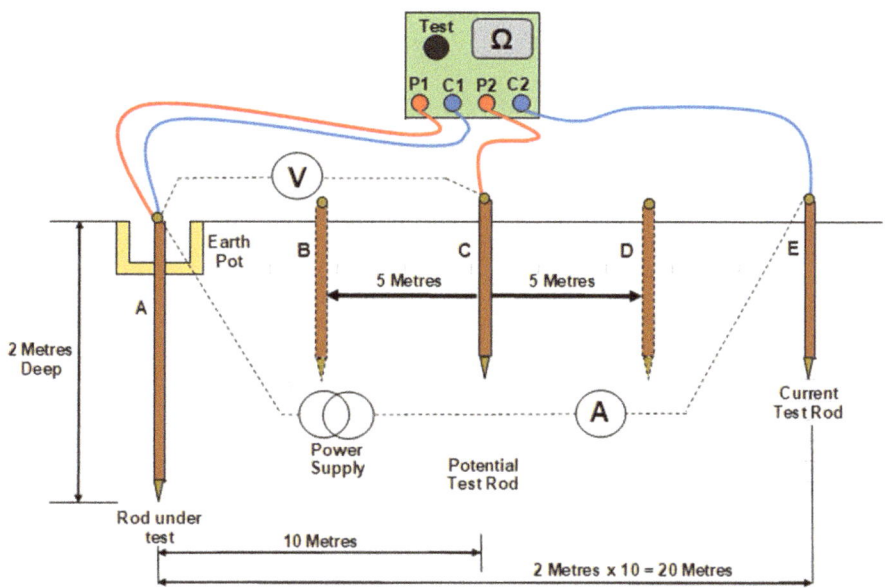

Basically what we start with is the electrode under test, which in this case is rod 'A' above. If the rod under test is, say, 2 metres deep then we must put the **Current Temporary Test Rod 'E'** around 10 times the depth away in the ground. About half way between 'A' and 'E' we then put **Test Rod 'C'** in the ground.

We connect the red and blue current and potential leads to the three rods as shown in the diagram above. We must ensure that the rods are outside the 'shells' of each other. **(Described under the Earth Rod Section.)** I have shown an 'Ammeter' and 'Voltmeter' with dotted lines to them to let you see what is going on inside of the Test Instrument which, using Ohm's Law, changes the reading to ohms for you on its screen.

We are now in a position to carry out the test, so we push the test button on the instrument and get a reading of **80Ω**. We now move test rod 'C' 5 metres nearer to the rod under test 'A' from its first position and get **79Ω** and now move test rod 'C' 5 metres further away from the rod under test 'A' and get, say, **84Ω**. These readings are around the same area so we can find the average of the three readings – **80Ω + 79Ω + 84Ω = 243Ω** divided by **3** comes to **81Ω**, this being our reading.

Now if our readings were still around the same area, but were something like **200Ω + 198Ω + 220Ω = 618** divided by **3** which comes to **206Ω**, then this would be far too high as our reading should have come out under **100Ω or better**, so either make **the earth rod under test** deeper by adding another rod on top or put another rod, joined to the first, in parallel to give more surface area.

If the readings were something like **80Ω, 200Ω and, say, 430Ω** which are not in the same area, then it is likely that **the test probes** are at fault so retest putting the test probes in different positions or deeper and see if the readings change for the better.

1) How often do you carry out testing of Earth Electrodes?

2) What Test Instrument do you use? Have you considered a Clamp Meter?

3) Have you a Procedure drawn up with expected readings?

Earth Clamp Meter:

The earth electrodes can be tested in several different ways these days. In my day it was by using a 'Bridge' Megger i.e. using a **Wheatstone Bridge** to find the earth rod resistance. Now the earth rod resistance can be found by using an **Earth Clamp Meter**. I have used one of these many times, saves an enormous amount of man-hours.

This of course is no ordinary clamp meter and just tests a local earth loop and nothing to do with star points of distribution transformers.

Although the manufacturers will of course not explain in detail how the instrument actually works we might guess.

All instruments measure current. We might call it a voltmeter or an ohm meter, but in the end it measures current. I can prove this by going to the old analogue Avo Meters. By the changeover switch we could measure amps, volts, ohms etc., but I have not changed the instrument. All I have done is changed the impedance or switched in diodes to go from AC to DC so that the moving coil meter could read AC.

I believe that this instrument is no different. The instrument reads current and changes it on the scale for you so that it appears to read ohms.

Because the manufacturers will not explain how it works, I will explain how I presume that it carries out the ohm measurement as follows: When the Instrument is first switched on an intermittent 'beep' can be heard, which I believe to be a signal generator which induces a voltage into the local earth loop that is to be measured, so we have the 'volts'. The rod/ground resistance is there and it is the constant that we want to measure, so we have the 'ohms'. So all we need to do now is measure the 'current' in the loop and by using Ohm's Law the instrument will recalibrate the scale from a 'current' reading to ohms for you.

The example on the right shows the Earth Loop that is being tested. The beauty of this earthing clamp meter is that the rod does not need to be disconnected and the test can be completed in a matter of minutes. Testing the conventional way requires the rod to be disconnected and electrodes hammered into the ground at a certain distance away, and usually takes around 25 minutes and a ground amenity test and permit.

1) Have you used an Earth Clamp Meter?

2) If 'YES' did you realise how much easier it was?

3) Do you realise how many man-hours it saves?

Earthing Drawing Documentation & Drawings:

FORM: ET 8						IS Industries Ltd.					
						Earth Pot Condition & Test Results					
Date of Test:	14/01/2021	Name of Tester:		I. Staff	Signature:	*Ian E Staff*			Plant:	P12	No. of Rods: 4
Earth Pit Number:	Location on Plant:	Type of Earth Pot:	Condition of Earth Pot:	Condition of connection?	Condition of Earth Cable?	Painted Green & Yellow?	Clamp Meter Reading?	Earth Bar Number:	Earth Bar Connection?	Clamp Meter Ser. No:	Comments:
EP1	SW Corner	Concrete	Good	Good	Good	Yes	0.9Ω	1	Good	12R679CM	
EP2	SE Corner	Concrete	Needs Changing	Good	Good	Yes	0.75Ω	6	Good	12R679CM	New Concrete Earth Pot required
EP3	NW Corner	Concrete	Good	Good	Good	Needs Re-Painting	0.8Ω	3	Good	12R679CM	Earth Pit Top requires painting Green/Yellow
EP4	NE Corner	Concrete	Needs Changing	Corroded - Greased	Good	Yes	0.9Ω	4	Good	12R679CM	New Concrete Earth Pot required

Correct documentation plays a huge part in the maintenance of earthing systems. The forms have to be clear and user friendly. In this section we discuss what test forms and location diagrams may be required to carry out testing of a plant earthing system.

It is important that the documentation is completed correctly as this will give the Engineer an idea of what is to be done in the future.

On the form the information required might be as follows:

1 – Earth Pot:- Pot number: Every Earth Pot should have a unique number.

2 – Location:- Exact location as sometimes old earth pots are difficult to find.

3 – Type of Earth Pot:- Concrete, Plastic.

4 – Condition of Earth Pot:- Cannot open, concrete broken etc.

5 – Condition of the Connection:- Where the Earth cable is connected.

6 – Condition of Earth Cable:- Condition of the cable going to the Earth Bar.

7 – Is the Earth Pot Painted Green and Yellow?:- Applicable? Company Policy.

8 – What is the reading from the Test Instrument?:- In Ωs.

9 – What is the number of the Earth Bar that the Earth Pit is connected to?

10 – What is the condition of the Earth Bar Connection?

11 – What is the earth Tester Instrument Serial Number?

12 – Comments:- Any incorrect readings or broken Earth Pits etc.

Also discussed are the drawing symbols that you may come across on electrical drawings associated with the plant.

See test form and location drawing on next two pages:

FORM: ET 8

IS Industries Ltd.
Earth Pot Condition & Test Results

Date of Test: 14/01/2021 Name of Tester: I. Staff Signature: *Ian E Staff* Plant P12 No. of Rods: 4

Earth Pit Number:	Location on Plant:	Type of Earth Pot:	Condition of Earth Pot:	Condition of connection?	Condition of Earth Cable?	Painted Green & Yellow?	Clamp Meter Reading?	Earth Bar Number:	Earth Bar Connection?	Clamp Meter Ser. No:	Comments:
EP1	SW Corner	Concrete	Good	Good	Good	Yes	0.9Ω	1	Good	12R679CM	
EP2	SE Corner	Concrete	Needs Changing	Good	Good	Yes	0.75Ω	6	Good	12R679CM	New Concrete Earth Pot required
EP3	NW Corner	Concrete	Good	Good	Good	Needs Re-Painting	0.8Ω	3	Good	12R679CM	Earth Pit Top requires painting Green/Yellow
EP4	NE Corner	Concrete	Needs Changing	Corroded - Greased	Good	Yes	0.9Ω	4	Good	12R679CM	New Concrete Earth Pot required

Location Drawings such as this one would have to be completed and attached to the Earth Testing Form for every plant so that Technicians can easily locate the required Earthing Pots and Earth Bars.

IS Industries Ltd.
P12 Ground Floor

Earthing Pots and Earth Bar locations:

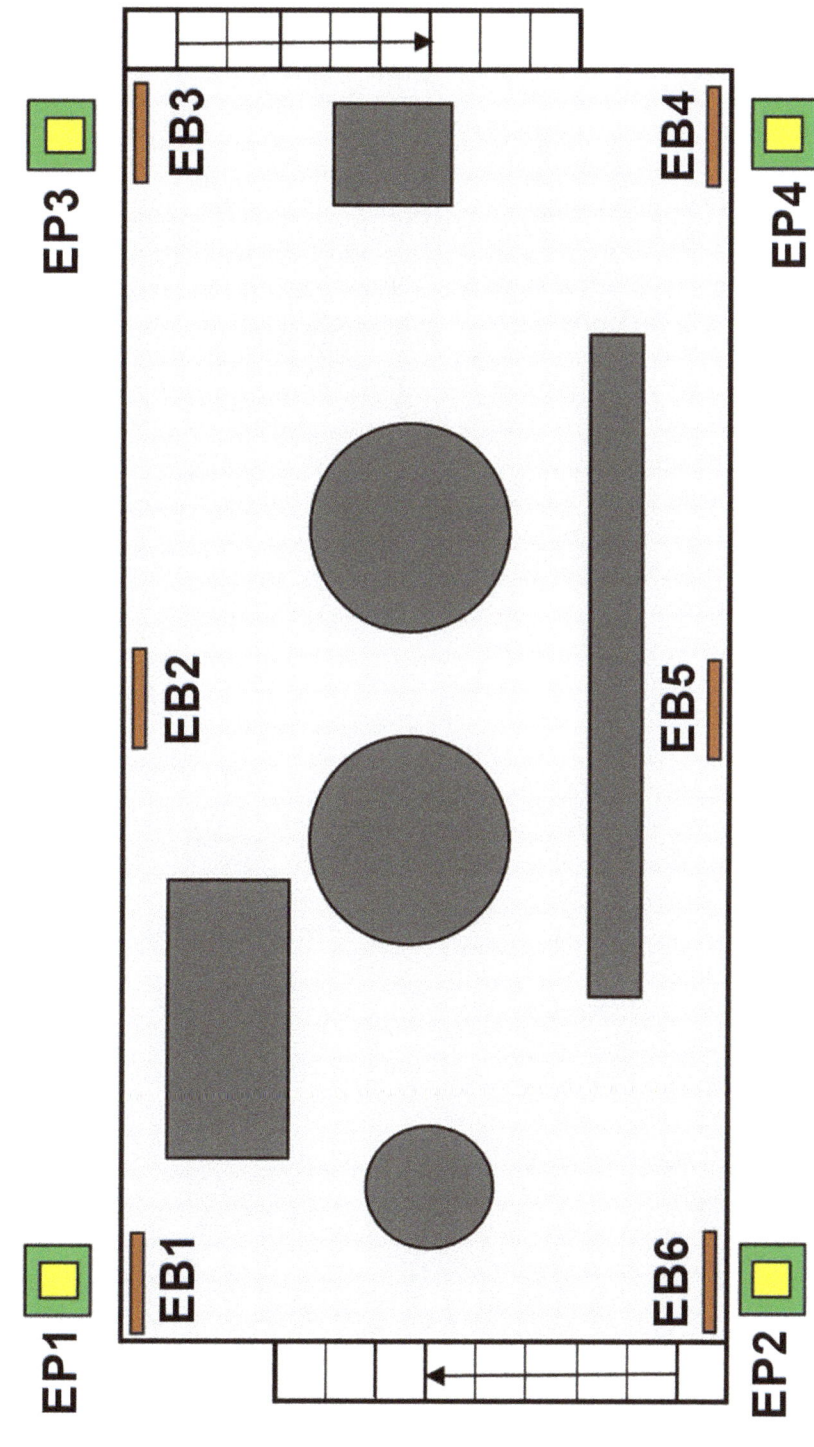

Location Drawings such as this one would have to be completed and attached to the Earth Testing Form for every plant so that Technicians can easily locate the required Earthing Pots and Earth Bars.

Earth Loop Impedance TN-S:

To find the **'Earth Loop Impedance'** we utilise the 'Live' and 'Earth' in, say, the socket outlet. So the earth loop that we are talking about is from the socket outlet through the earth connections in the dis-board, through the MV panel and circuit breaker, up the star point earth of the distribution transformer, through the phase winding and back through the MV panel, through the fuse/circuit breaker **(in yellow below)** in the dis-board and back to the Live in the socket outlet. This loop is marked with black arrows in the diagram below.

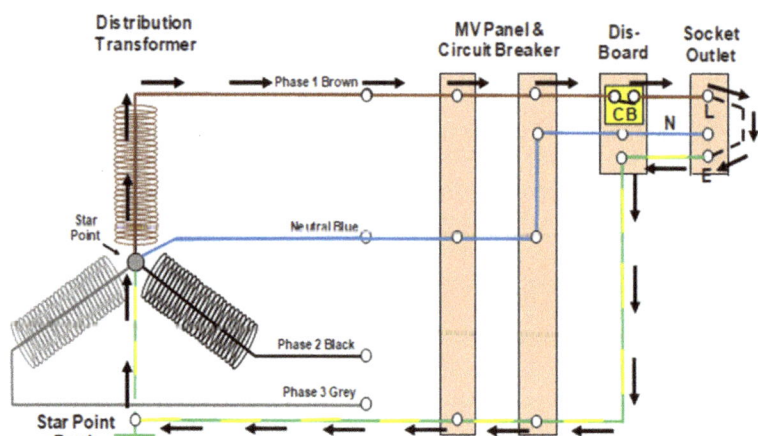

So what are we trying to achieve? Well we know the voltage which is **230V**, but now we need to know the 'Impedance' and by using Ohm's Law we can find out what current would flow in the event of a fault, which of course must be enough to blow the fuse or trip the circuit breaker or blow the fuse in the dis-board! Also if the earth wire is only fastened by a couple of strands the high current will blow it off.

When we are deciding how to work this impedance out what are we looking for? Well firstly anything that is external to the building is **Ze** ('e' for external). When we consider the impedance of the entire system we call this **Zs** ('s' for system). So what Zs impedance are we looking for on our **TN-S** Earthing System? Well anything around **0.8Ω**, so looking at the current in this case and applying Ohm's Law V/R, we would be looking at **230V/0.8Ω = 288 amps.** As well as the external impedance (**Ze**) we also have the internal equipment/wiring from the external to the socket and back to consider and usually this is measured in Resistance as no coils are involved. So our formula may look like: **Zs (Whole System) = Ze (External) + R1 + R2 (Internal).**

We need to know that the protection will protect us. I will explain in domestic terms. What we have to work out now is: Would our 'Loop Current' of **288 amps** trip the circuit breaker in the dis-board? Let us say that we have a 32-amp circuit breaker. The current needed to trip it is as follows:

Circuit Breaker Type B – 3 to 5 times rated current to trip - **32 x 5 = 160 amps**

Circuit Breaker Type C – 5 to 10 times rated current to trip - **32 x 10 = 320 amps**

Circuit Breaker Type D – 10 to 20 times rated current to trip - **32 x 20 = 640 amps**

So back to our loop current of **288 amps**, this would be more than enough to trip a circuit breaker **Type B – 160 amps**. We could not use circuit breakers type C and D as there would not be enough current flowing to trip them.

The maximum current to earth required to operate the circuit breaker/fuse is called the **Prospective Fault Current (PFC)**. This should be tested at point **Ze (external)** with the main switch off. This is where the **'Prospective Fault Current'** would be high because of the low resistance, say, **230V/0.2Ω = 1150 amps**. Further down the line towards the transformer the current would decrease as the **Zs** impedance increased.

1) Did you realise why Earth Loop Impedance is so important?

Earth Loop Impedance TT:

To find the 'Earth Loop Impedance' we utilise the 'Live' and 'Earth' in, say, the socket outlet. So the earth loop that we are talking about is from the socket outlet through the earth connections in the dis-board, through the MV panel and circuit breaker, down the panel earth rod through the ground and up the star point earth of the distribution transformer, through the phase winding and back through the MV panel, through the fuse/circuit breaker **(in yellow below)** in the dis-board and back to the live in the socket outlet. This loop is marked with black arrows in the diagram below.

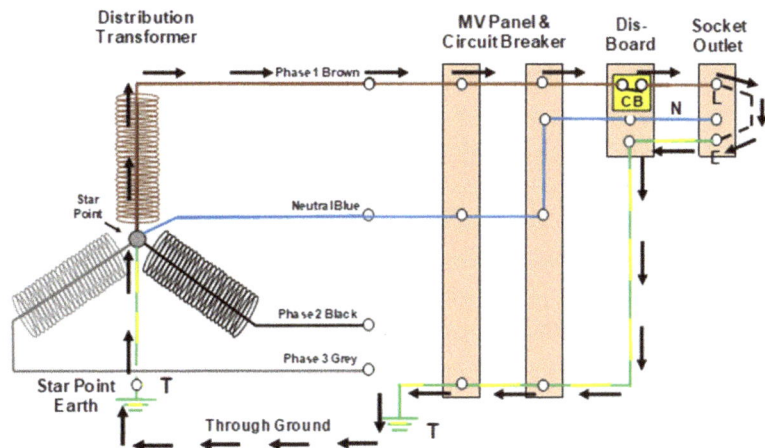

This is a TT (Terre Terre or Rod Rod) Earthing System. So what are we trying to achieve? Well we know the voltage which is **230V**, but now we need to know the Impedance and by using Ohm's Law we can find out what current would flow in the event of a fault, which of course must be enough to blow the fuse or trip the circuit breaker in the dis-board.

When we are deciding how to work this impedance out what are we looking for? Well firstly anything that is external to the building is **Ze** ('e' for external).

When we consider the impedance of the entire system we then call this **Zs** ('s' for system). So what **Zs** impedance are we looking for on our **TT** Earthing System? Well anything around **20–50Ω,** so looking at the current in this case applying Ohm's Law V/R we would be looking at **230V/50Ω = 4.6 amps.**

As well as the 'external' impedance (**Ze**) we also have the internal equipment/wiring from the external to the socket and back to consider and usually this is measured in resistance as no coils are involved. So our formula may look like. **Zs (Whole System) = Ze (External) + R1 + R2 (Internal).**

What we have to work out now is: Would our **Loop Current** of **4.6 amps** trip the circuit breaker in the dis-board? Let us say that we have a 32-amp circuit breaker. The current needed to trip it is as follows:

Circuit Breaker Type B – 3 to 5 times rated current to trip – **32 x 5 = 160 amps**

Circuit Breaker Type C – 5 to 10 times rated current to trip – **32 x 10 = 320 amps**

Circuit Breaker Type D – 10 to 20 times rated current to trip – **32 x 20 = 640 amps**

So back to our loop current of **4.6 amps, THIS WOULD NOT BE ENOUGH** to trip the lowest circuit breaker **Type B – 160 amps**. We would therefore have to supplement the trip current by using **Earth Leakage Units (ELU)** or **Residual Current Devices (RCD)** with milliamp trips.

I have given this example using **domestic** circuit breakers to demonstrate the importance of the earth impedance value. Decisions to use ELUs or RCDs is sometimes vital rather than just a good idea.

1) Did you realise why ELUs sometimes back up a TT System?

Testing Motor Earth Path:

Motor earth path tests are carried out by many companies to ensure that the motor actually has a good earth. Sometimes the earth to the motor is provided by a 4mm bonding wire and sometimes it relies on the gland.

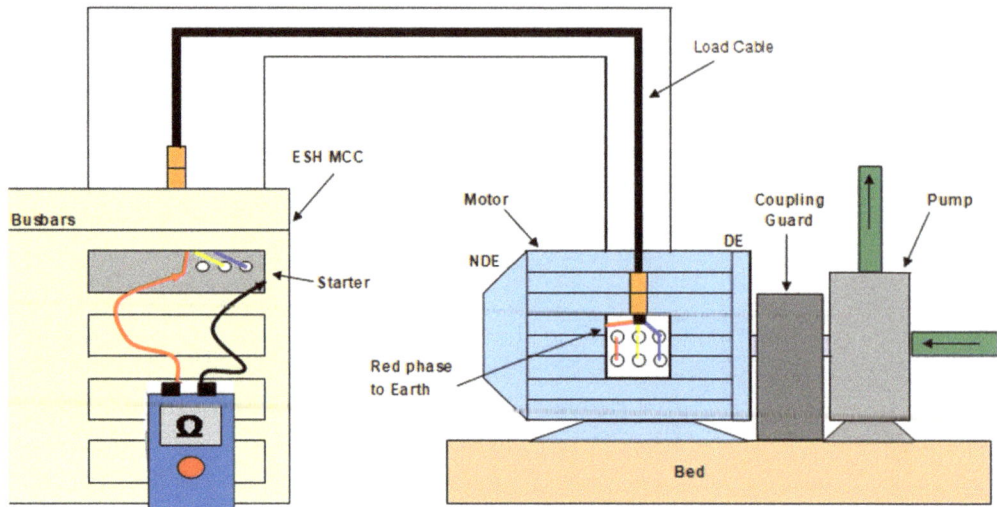

Earth Path Test from electrical equipment on site, which could be a motor or a light fitting etc. The reading should be in ohms. In the above diagram we are using one phase of the load cable to the motor, similar to a loop impedance test, except here you are just testing the loop resistance from the starter to the motor and no high currents.

The phase is connected to earth out in the field and checked from earth to the phase that you have used back at the starter in the MCC. The earth path reading should be around 0.8 ohms.

Some companies do this a slightly different way. They connect, say, the red and blue together at the motor and put the meter between red and blue at the starter and get a reading in ohms. Half this reading will give the resistance of one phase wire, say red. Then they do the earth path test as described above, and take off the reading of the core to provide a pure earth path. As the readings are so small, in my opinion this is a waste of time. I have taken the cable into the top of the motor terminal block for ease of understanding. Normally, of course, it would be in the bottom.

We have to ensure that any earth protection device, which is protecting the motor inside of the starter unit, will actually work if there was to be an earth fault on the motor. This may be in the form of fuses or circuit breakers. This would not include a core balance CT as with this unit the trip would occur to an imbalance of the phases, possibly due to an earth fault.

Just remember, if the motor is in a Zoned Area any sparks from this test will not be intrinsically safe and neither will the instrument. A Gas Free Certificate should be considered, depending upon Company Policy.

1) Do you carry out this test as part of a Detailed Inspection of Equipment?

Earth Leakage Units and RCDs:

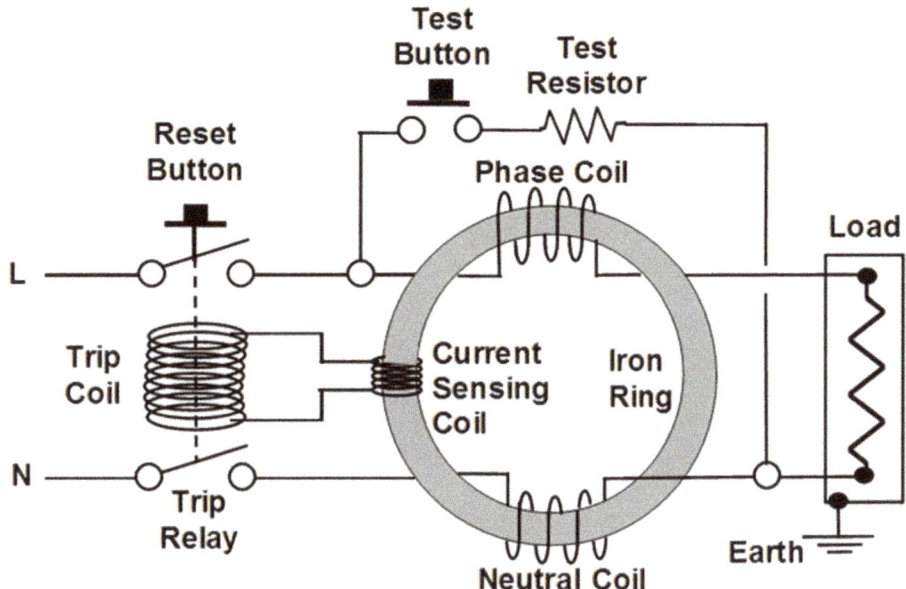

Residual Current Devices (**RCDs**) are different from Earth Leakage Units (**ELUs**) but they both achieve the same objective which is to trip the circuit relay if a certain milliamp leakage develops which may be a threat to human life. Sometimes RCDs are referred to as Residual Current Circuit Breakers (**RCCDs**) and sometimes ELUs are referred to as Earth Leakage Circuit Breakers (**ELCB**). Remember that these devices are operated efficiently on '**leakage**' current not a dead short circuit. Other devices such as fuses and circuit breakers look after these although this relay may trip as well.

It takes around 33 milliamps (33 thousandths of 1 amp) to kill Mr or Ms Average so these devices usually operate at 30MA or below. If the milliamp trips are too low there may be nuisance tripping. In the past I have known schools to be as low as 12MA, but these days I would imagine that a 20MA Unit would suffice. Later I will explain the '**voltage**'-operated earth leakage circuit breaker that was more popular in the 1980s before the IEE Regulations removed it. We can look at some of the problems that arose. These were abbreviated to '**VOCB's**.

Current-operated earth leakage units are used as back up where the earth loop impedance current was, say 4.5 amps on a TT system, or where there was an impedance of 5Ω. The low loop current would not be enough to trip the protection circuit breaker so some other protection must be used. Also on some TN-S systems with a higher loop current, certain value circuit breakers would trip, but some may not so a ELU back up is required.

For nuisance tripping, unplug or switch off every device/piece of equipment in the circuit and see if the unit will reset. Switch devices on one by one until there is a trip and you have found the problem. If this fails the fault could be in the circuit wiring or a tired earth leakage circuit breaker.

If we use a domestic example i.e. your lawnmower, these days the blade travels so fast that when it cuts through the cable the short circuit, caused by the blade on the live and neutral, is on for just a few microseconds so the fuse does not get chance to blow. People will for some reason pick up the live cable and, if not protected by an RCD or ELU, get electrocuted. The test button is just to show that it is working, but not a true test.

1) How often do you test your RCDs/ELUs?

2) Did you realise that the RCDs and ELUs operate differently?

Residual Current Circuit Breaker (RCCB):

This unit is sometimes called as above a Residual Current Circuit Breaker **(RCCB)** and sometimes a Residual Current Device **(RCD)**. Sometimes these devices get mixed up with an Earth Leakage Unit **(ELU)**, especially on the internet, and although they both achieve the same objective, and are tested the same, they are different in the way that they work.

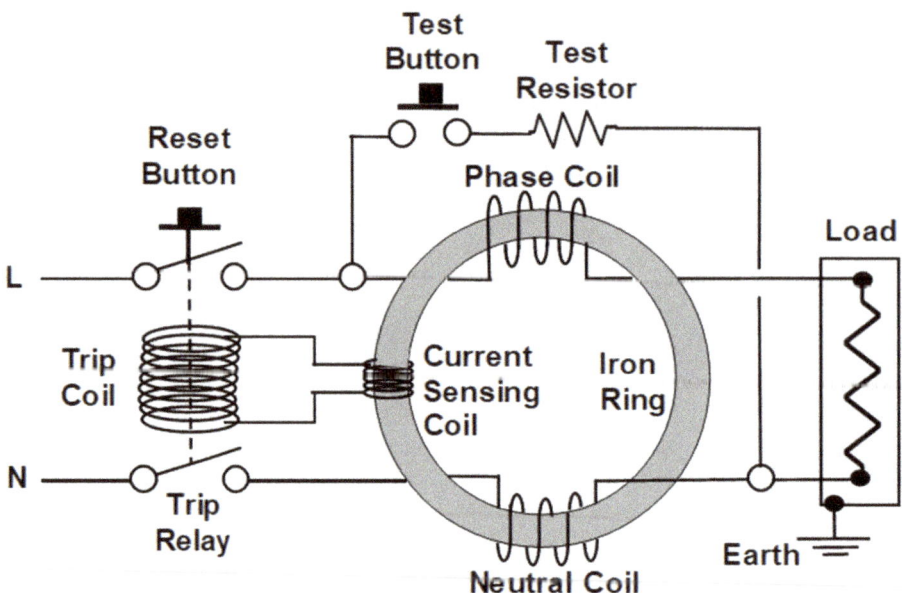

In our Hazardous Area we will definitely have some socket outlets which are 230V and 415V. In a factory these will probably be protected by an Earth Leakage Unit **(ELU)** which, as I have stated above, is slightly different in the way it works to our **RCD** and is explained later.

Let us stick for the moment to our RCDs. As you can see in the diagram above, it consists of many parts. This device relies on what comes down the live or phase wire and goes back down the neutral wire. As you can see in the diagram above, the device consists of an **Iron Ring** with three coils wrapped around it i.e. the Phase or Live Coil, the Neutral Coil and the Current Sensing Coil.

Let us first look at the live and neutral coils. Wrap a coil around an iron ring and put power onto it and what happens? It turns into an electromagnet. If we wrap the phase coil one way round the ring and the neutral coil the other way, as above, when the circuit breaker is reset the live goes through the phase coil, through the load and back the other way round through the neutral coil and they cancel each other out so the iron ring **DOES NOT** become a magnet, so the current sensing coil is dormant.

If however there is a milliamp earth fault and, at home, that could be someone picking up a cut lawnmower lead that is live, what is going down the live and through the load is not all going back through the neutral coil. Some is returning through the person to earth, Now the two coils are imbalanced so the iron ring **DOES** become a magnet oscillating at 50Hz. The lines of force from the magnet cut through the search coil causing an EMF to flow through the **Current Sensing Coil** and trip the relay.

So that a rough test can be carried out, there is a Test Button going through a test resistor which is there so that the test does not put a dead short circuit onto the wiring. The unit should always be tested with an 'ELU' tester and not rely solely on the test button. There is also a reset button.

1) What milliamp setting are the ELUs on your plant?

Earth Leakage Unit (ELU):

This unit is sometimes called as above an Earth Leakage Circuit Breaker (**ELCB**) or an Earth Leakage Unit (**ELU**) or Core Balance Unit (**CBU**). These devices get mixed up with a Residual Current Device (**RCD**), especially on the internet and although they both achieve the same objective, and are tested the same, they are different in the way that they work.

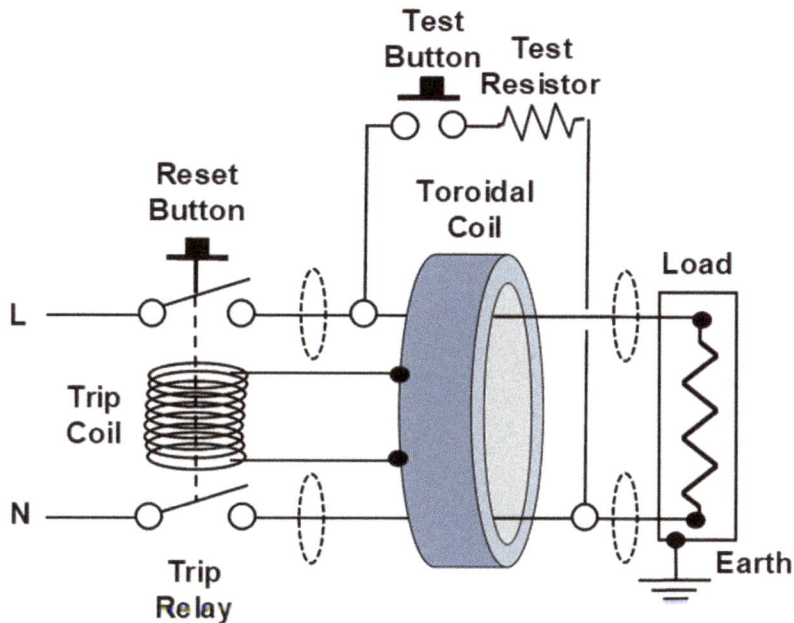

In our Hazardous Area we will definitely have some socket outlets which are 230V and 415V. In a factory these will probably be protected by a **Core Balance Unit (CBU)**, above, and which I have explained earlier is slightly different in the way it works to our **RCD**. What we have here consists mainly of a toroidal coil where the live and neutral conductors pass through it. The coil is much like a **current transformer** using the 230V/415V cores as a primary and the coil the secondary. The conductors have magnetic fields (the dotted lines around the conductors, above) around them which are oscillating at 50HZ. Because the live core field cancels out the neutral core field the toroidal coil is dormant.

If however there is an earth fault or someone touches the live part of the load then what is coming down the live is not all going back down the neutral as some of the current is returning down through earth. In this case the strengths of the magnetic fields become imbalanced causing an EMF to flow round the toroidal coil and out to the trip coil causing the trip relay to trip the circuit breaker.

They are usually reset by taking the red cut out switch, pushing right down to the bottom to arm it and then lifting it up to the top. These units are usually set to trip at **30 milliamps** in a set number of **microseconds**. Remember these units are usually to protect against milliamp earth leakage not short circuits. In schools the units might be set at 20 milliamps. Too low and there may be nuisance tripping.

One thing that must NEVER happen here which could be a possibility is the earth wire of say a 3-core cable passing through the coil as well as live and neutral, or the unit will not trip, as this would balance the cores EARTH FAULT OR NOT. The trip switch is usually the reset. The consumer must push the switch completely to the bottom to latch it and then up to the 'on' position. There is also a test resistor and test button for a rough test, but the unit should always be tested by an ELU Tester.

Voltage-Operated Earth Leakage Unit:

Back around the 1980s another type of earth leakage unit was used on older installations and that was a voltage-operated earth leakage unit. There were problems with this type of unit. Older editions of the IEE Regulations banned this type of earth fault protection. Let us see how it works.

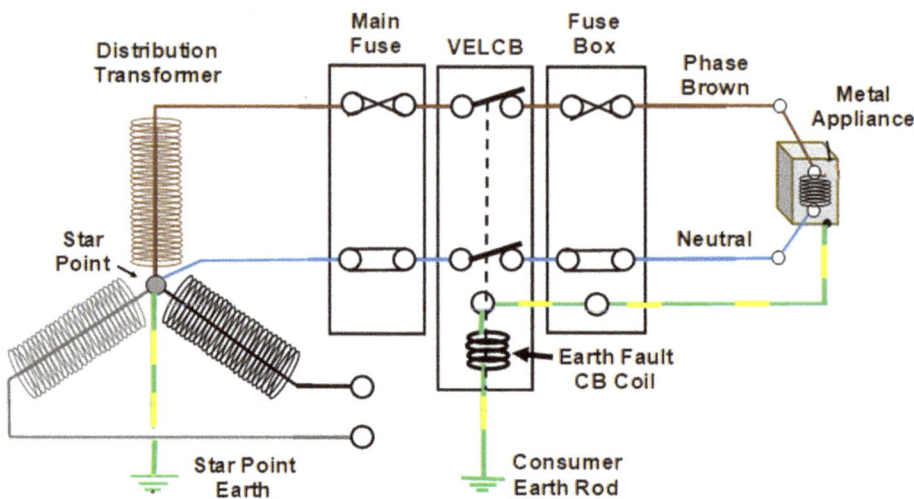

These units of course have to work on a TT earthing system and, as I mentioned earlier, they are an older earth protection unit so older ones may still be in use, but modern units with ELUs or RCDs will work more with **current** than with **voltage**.

Let us say that we get an earth fault on our metal appliance on the right hand side of the above diagram. The path of the voltage would be from the case through the earth point in the fuse box through the coil in the Voltage Earth Leakage Circuit Breaker (**VELCB**) and down to the rod.

The action of opening the double pole circuit breaker contacts actually **CLOSES** another contact, which directly shorts out the coil and makes a direct circuit for the earth fault down to the rod. There is a test button to a resistor available so the mechanism can be tested to see if it works.

So voltage passing through the coil to the rod operates the unit double pole circuit breaker and trips the unit. Now on the face of it this seems a very efficient piece of equipment, but I am sure that you can see glaring problems with this type of unit:

1 – If someone was working on the appliance whilst it was live and they got in contact with the voltage THEY might be a better connection to earth than the unit earth coil and get electrocuted. So the unit only protects equipment faults to earth.

2 – Say the wire was chopped off our appliance, but remained live. If you picked it up and got in contact with the live wire then there would be no path back to the coil/rod so the voltage/current would pass through you.

3 – If for any reason the rod went very high impedance or the connection to the rod was lost then the unit would not work and the appliance case could become live in the event of an earth fault.

    1)   Have you any Voltage Operated Earth Leakage Units? (I doubt it)

    2)   Were you aware how dangerous they could be?

Motor Earth Leakage Protection:

What type of earthing system do we have in our Hazardous Area and what earth leakage protection are we going to have for our motors?

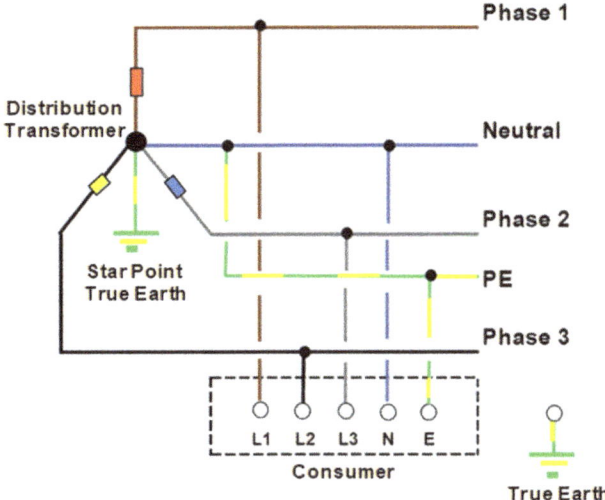

Firstly let me say that **TN-C & TNC-S** systems where the earth and neutral are combined are **NOT ALLOWED** in Hazardous Areas! Diagram to left.

With these systems with the earth **'Combined'** there could come a time at the consumer end that the potential of the earth provided by the system and true earth are different.

In this case there could be, say, a motor on site fed from a switch room which is different to the plant around it, which would, of course, be dangerous.

Earth faults inside a motor can very badly damage the windings with the high current produced. The above systems cannot be installed in our Hazardous Area and it is unlikely that an IT system is fitted where there is a high impedance in the distribution transformer star point earth to stop these high currents. The system that is installed in your Hazardous Area is probably a TN-S where the earth is **'Separate'** or a TT system using earth rods (Terre Terre).

So how do we protect our motors? Well just after the isolator we can take our three phases through the contactor and out to the motor through a core balance unit.

It might be a decision by the electrical engineer in charge that core balance is fitted to, say, every motor over 5KW. The core balance unit would detect the imbalance caused by the earth fault before any serious harm was done.

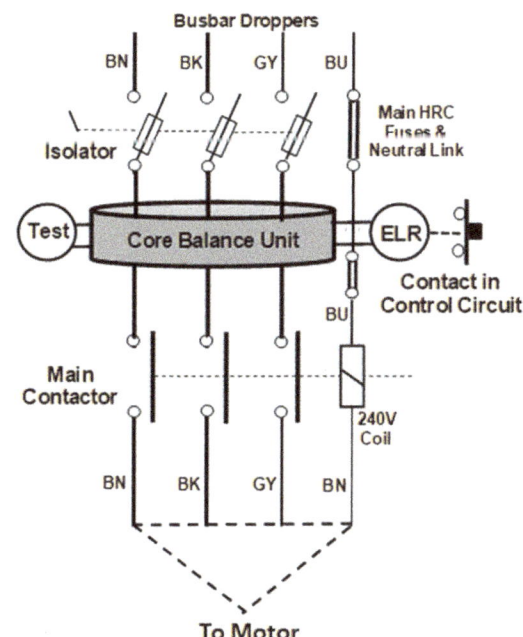

A light on the starter front would indicate that the earth leakage relay had in fact operated, and there would be a reset button to reset the system **WHICH MUST NOT BE RESET WITHOUT AN IR/BALANCE TEST ON THE MOTOR FROM THE STARTER.** There is usually a key operated test facility on the starter front which puts a resistance on to the system to make it test trip. Earth faults on the load cable will also operate the unit.

The core balance here may be set at around 5 amps which is motor protection, *not human.*

1) Do you have core balance on motors? If so, on what Maximum Size Motors?

Clean Earths:

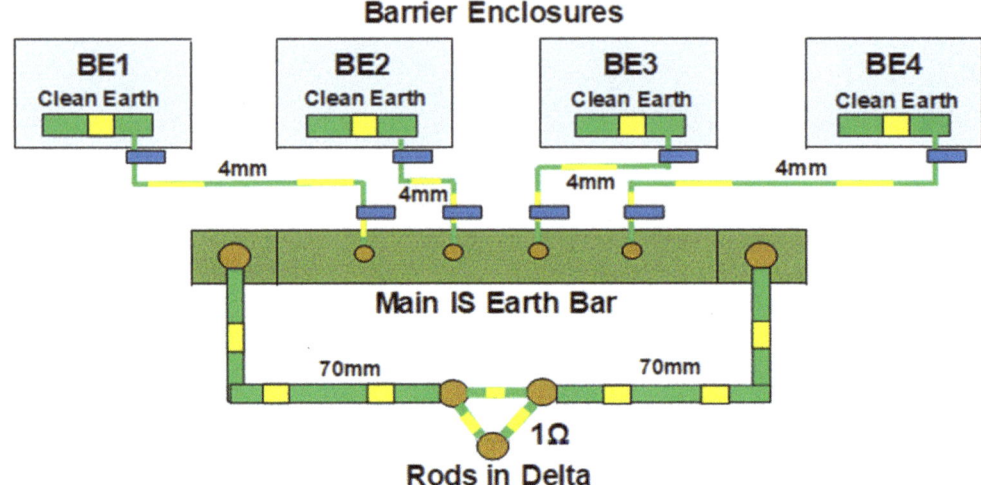

When it comes to Intrinsic Safety then **'Clean'** earths are required. These are earths usually for Zener Diodes or Screens that require an earth bar going to independent earth rods with impedance of **1Ω or better**. I must stress again that **'better'** might be quite difficult to achieve.

At the top is a diagram showing several barrier enclosures **(BE)** where their **SINGLE WIRE 4mm** clean earth feeds go to a huge earth rod in a noise free location which, in turn, uses **70mm** cables to go out to separate clean earth rods.

In many companies the 4mm cables coming from the barrier enclosures will be duplicated. It is possible, if there are two clean earths from the barrier enclosures, to use **2 x 1.5mm** earth wires.

Our clean earth must be totally free of electrical 'noise' and must be free of dangerous invading voltages from other electrical systems. Clean earth cables must not be run together in a cluster with other earth cables such as power earths which could contain electrical noise.

Also, I say independent, as we do not want our Intrinsically Safe earth wire going to a power earth bar, which could be **5Ω** or worse, and then to the structure lightning protection earth which could be **10Ω**.

It is possible that using a totally separate system for the **'clean'** earth might lead to a potential difference between this and the power earth, which is one of the reasons that a TNC-S system may not be used in a hazardous area.

It might be an experiment, in a safe area or Zone 2 with a Gas Free Certificate, to get a multi-meter and test the potential between your clean and dirty earths.

Intrinsically Safe systems may not be the only systems using clean earths as telecommunications may also require them.

1) Do all Clean Earths go to their own independent rods on your plant?

2) Have you up-to-date Hook-up/Loop Diagrams showing earthing?

3) Do all earths' numbers match those shown in cabinet at the main Clean Earth Bar?

Barrier Box Clean Earth Bar:

Firstly let me explain that barriers in general limit the energy from the power supply, which is usually in the non-hazardous area, to the field which is our Hazardous Area. There are two main types of barrier, namely Zener and Galvanic. How these actually function is shown in my other book: 'Hazardous Areas for Technicians'. The diagram below shows several Zener Barriers in a barrier box.

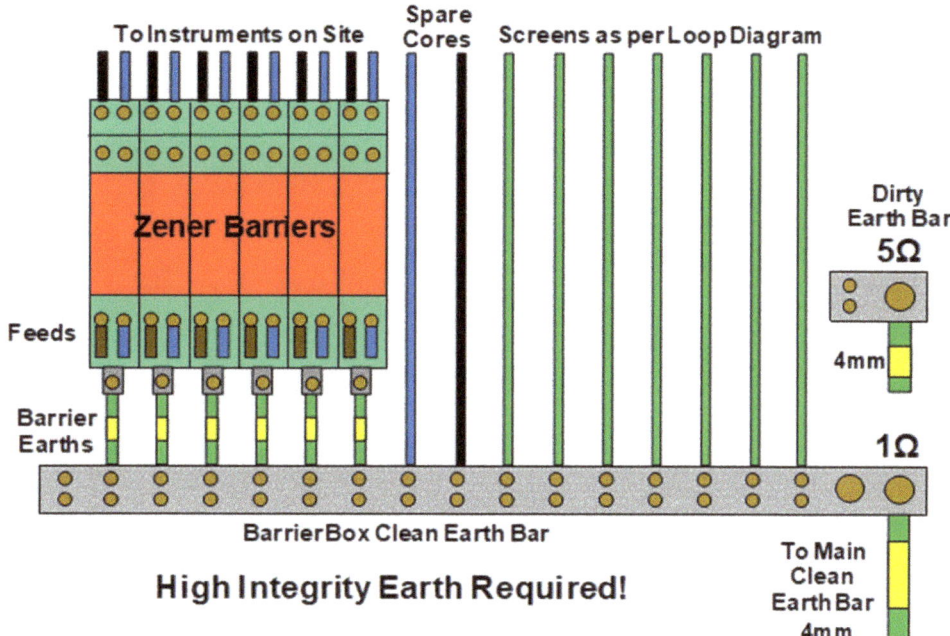

High Integrity Earth Required!

The barrier box above has its own clean earth bar which is fed, usually from a main clean earth bar, by a **4mm** earth cable. (If there are two feeds then they can be **1.5mm** each.) This clean earth bar must have a high integrity earth of **1Ω**. There is also a dirty earth bar, again with a 4mm feed, which could come from a general building earth bar and could be **5Ω**.

Going to the earth bar in this case are spare cores, from multicores concerned with loops in this configuration, and in the example diagram above screens (which are green not green and yellow) from the blue and black outgoing instrument cables to site. (It must be remembered that screens must be connected as per 'hook up' or 'Loop' Diagram.) Most important, also going to this earth bar, are 'Earths' going to the Zener Barriers, as without these the Zener Barriers will not be able to perform their safety job. The worrying part is that the Loop will still function without the Zener Barrier earths.

So let us look at this **1Ω** earth. TNC earthing systems are not suitable for hazardous areas and certainly not here! IT systems may also not be suitable in this case and only TN-S or TT earthing systems should be used. Standard earths could be **5Ω** going up to **10Ω** for the plant steelwork so we do not want our IS earthing to go in this direction. These high integrity earth bars going to separate IS earth rods from the main earthing system ensure that any fault currents from the **5Ω** system cannot be directed down to IS equipment. Also we are dealing with such small voltages and currents that we need an earthing system with as little resistance as possible.

1) Do you have Zener Barriers on your plant?

2) If 'YES' are the Earths and Screens connected correctly?

Barrier Box Clean Earth Bar:

Firstly, let me explain that barriers in general limit the energy from the power supply, which is usually in the non-hazardous area, to the field which is our Hazardous Area. There are two main types of barrier, namely Zener and Galvanic. How these actually function is shown in my other book: 'Hazardous Areas for Technicians'. The diagram below shows several Galvanic Barriers in a barrier box.

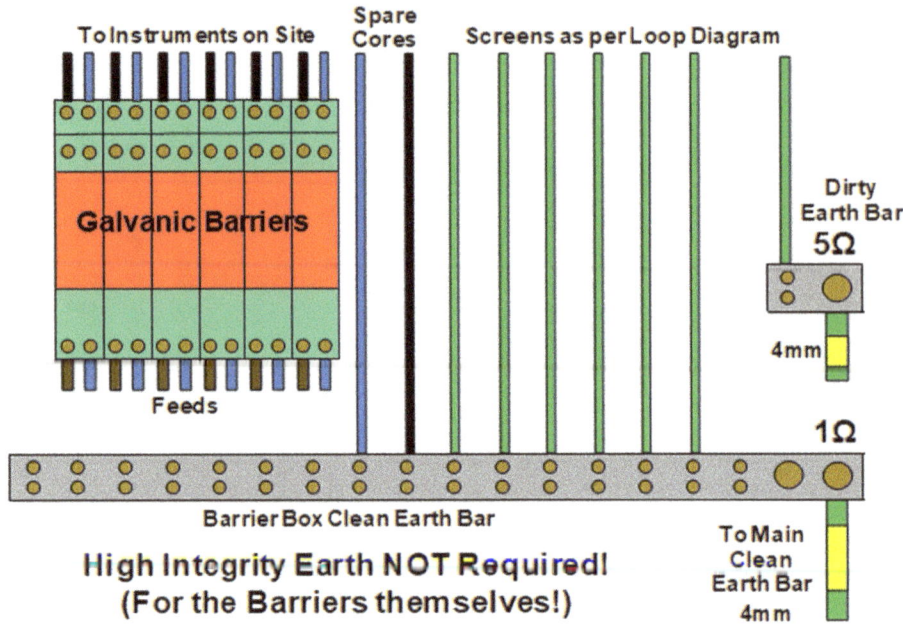

The barrier box above has its own clean earth bar which is fed, usually from a main clean earth bar, by a **4mm** earth cable. (If there are two feeds then they can be **1.5mm** each.) This clean earth bar must have a high integrity earth of **1Ω**. There is also a dirty earth bar, again with a 4mm feed, which could come from a general building earth bar and could be **5Ω**.

Going to the earth bar in this case are spare cores, from multicores concerned with loops in this configuration, and in the example diagram above, screens (which are green not green and yellow) from the blue and black outgoing instrument cables to site. (It must be remembered that screens must be connected as per 'hook up' or 'Loop' Diagram.) The Galvanic Barriers themselves do not require a high integrity earth as the way that this barrier functions is totally different from the Zener Barrier.

So let us look at this **1Ω** earth. TNC earthing systems are not suitable for hazardous areas and certainly not here! IT systems may also not be suitable in this case and only TN-S or TT earthing systems should be used. Standard earths could be **5Ω** going up to **10Ω** for the plant steelwork so we do not want our IS earthing to go in this direction.

These high integrity earth bars going to separate IS earth rods from the main earthing system ensure that any fault currents from the **5Ω** system cannot be directed down to IS equipment. Also we are dealing with such small voltages and currents that we need an earthing system with as little resistance as possible.

1) Do you have Galvanic Barriers on your plant?

2) If the answer is 'YES' are the Earths and Screens connected correctly?

Clean Earth Intrinsic Safety:

Below is a diagram where I have shown four Barrier Enclosures which could be large panels in the 'auxiliary' room of a control room. Each of these four enclosures has its own clean earth bar to which the barriers, screens and spare cores of multicores are connected. These are connected via a **4mm cable** to a huge 'Main' IS earth bar which could be under the control room floor and this goes out to the rods.

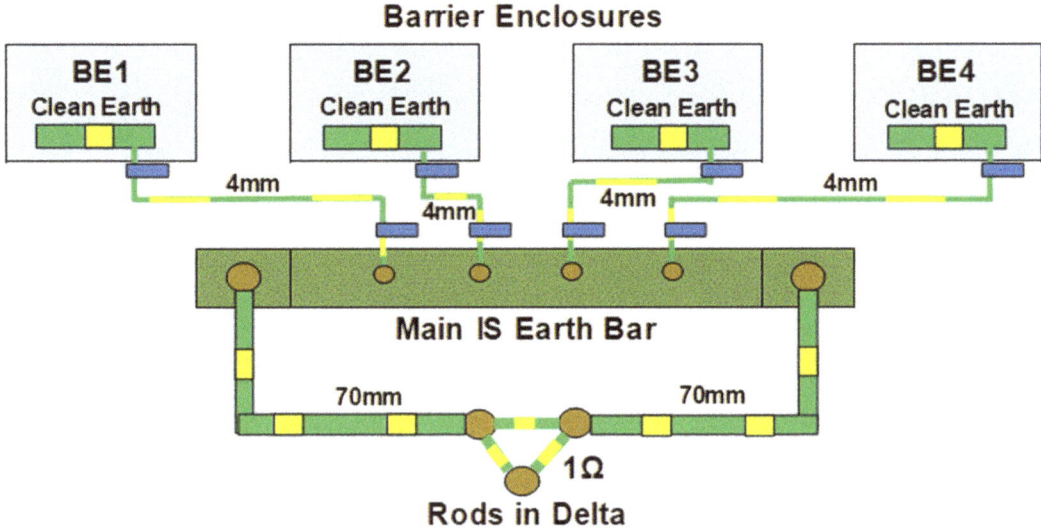

As mentioned above, the main IS earth bar will go out to the rods possibly on two 70mm cables. I have only drawn four barrier enclosures, but there may, in reality, be many more. The earth rods are on their own and separate from the power rods. There may be what we came to know on our factory as an earth cluster i.e. three rods joined up in delta to get the amount of copper in the ground to achieve minimum **1Ω**.

If there is only one earth cable in the barrier boxes, under no circumstances must it be removed for testing if the plant is running, as if the barriers are Zener then that would leave them unprotected.

It may be noticed in the diagram that all IS earths have blue tags on them. This is so that at the main earth bar, Instrument Technicians can identify which barrier enclosure or cabinet the cables came from.

We did prove that on our particular plant we could use the earth clamp meter to test individual earth cables without affecting the running of the plant, but this must be done with Engineer approval and a risk assessment. There may also be some trial and error before this is performed.

We keep calling them **'Clean'** earth bars, but the name really depends upon what the company policy is. In some companies these may be called **'instrument'** earths, **'IS'** earths, **'insulated'** earths or **'noiseless'** earths.

Clean Earth Rods (**1Ω**) should be approximately **2 metres** away from Power Earth Rods (**5Ω**). Some equipment generates what is called **'Common Mode Noise'** which, if connected to a Power Earth Bar, can affect other delicate equipment which is connected there.

1) Do you have a main IS Earth Bar and do you know where it is?

Lightning Protection:

Many people ask 'if the columns on a chemical factory are quite high and made of metal why do they need a lightning conductor?' The answer might be that most of the time they don't. The problems arise if they get hit and possibly damaged. Remember columns have vents, which are Zone 1 Areas, around them. Apart from structural damage, lightning could ignite the vented gases etc. The conductor provides a definite path. There does not have to be a very good resistance to earth from the lightning conductor, so long as there is one i.e. 10Ω would be sufficient.

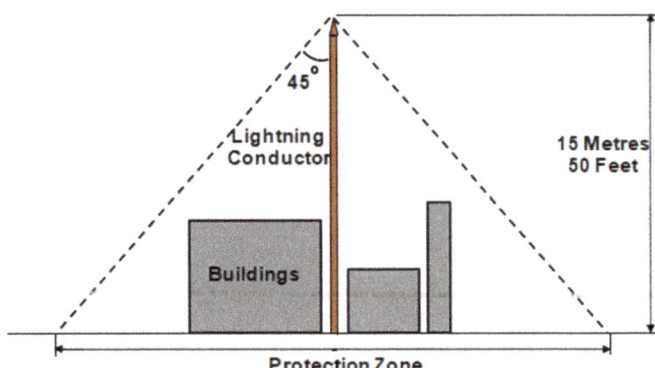

If we want to protect our chemical plant against lightning there are several calculations that must be considered for maximum safety. The following pages explain why we put the lightning conductors up and how they might protect our chemical plant.

There is no system that I know of that can guarantee 100% protection against lightning. All that we can do is ask experts and do our calculations, estimates and Risk Assessments. There are devices for surge protection which we will discuss later. Remember that you may have quite a large amount of sensitive equipment, for example Intrinsically Safe equipment, that may need protection from lightning surges because they are in Zone 0 and Zone 1 Areas.

Many people are under the impression that a lightning bolt has to actually hit you to cause death. I can confirm that if the bolt did hit you directly then you would most probably die. Opposite and on the next page I have demonstrated that the lightning does not have to hit you directly to harm you. All it has to do is hit the ground somewhere in your vicinity.

People also think that sheltering under a tree in a storm is safe, but this is not so. The moisture in the tree will conduct electricity to the ground and the ground potential will rise. The effect will be similar to my dartboard effect on the next page.

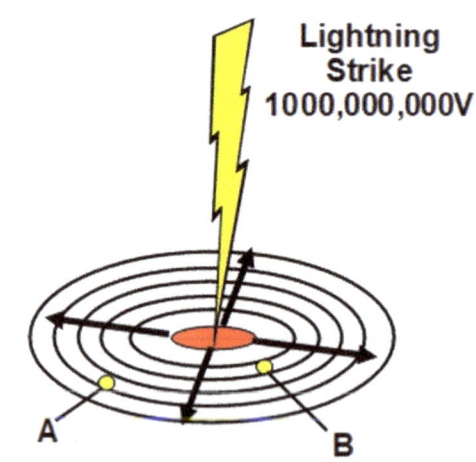

Another phenomenon that could happen to the tree is that, as the moisture turns to steam, the tree could explode quite violently.

1) Are there lightning conductors on tall structures on your plant?

2) Have conductor angles (above) been worked out for maximum protection?

3) How often have you known the plant get struck?

4) Have you got lightning 'surge' protection?

Lightning Protection Standards: IEC 62305, IEC 62561, EN 50164

We have come a long way since the 18th century when Benjamin Franklin flew his kite into the clouds, famously dangling a key from the silk thread, but thanks to him we got to know quite a lot about electrostatic charged clouds.

Facts: Lightning can be 100,000,000 volts and 20,000+ amps. The average flash is usually around four miles long. The longest flash ever recorded was allegedly 400+ miles long over Brazil, recorded in 2018 (don't ask me how they measured it!). The flash temperature is around 30,000 degrees Celsius (five times hotter than the sun's surface which is reportedly 5,500 degrees Celsius). The thickness of the arc is around 1.5-2" but it looks thicker because of the brightness. The flash speed is around **60,000 miles/second** whereas sound is around **761 miles/hour** so you will see the flash well before you hear the thunder. Ask quite a lot of people 'why do we put lightning conductors on tall buildings?' and they will say 'to attract the lightning'. So we have all that energy above floating past on a cloud, minding its own business and we want to put copper rods in the sky to attract it down here!

So how do we protect our plant? Well for a start putting lightning rods on the tallest structures down to an earthed rod around 10Ω impedance or better. If the structure did get hit, the rod would direct it via the shortest route to earth, hopefully decreasing the damage. Another phenomenon that happens in the event of a strike is that an arc ascends upwards from the rod to meet the descending arc, hopefully taking a lot of the energy out of the strike and directing it to the ground.

Now if we look at the whole of the plant lightning system, what is happening all of the time during a storm is that the rods are constantly emitting positive (+) ions, draining the charge out of the storm clouds before any lightning strike happens, thus lessening the chance of there actually being a strike in the first place.

It is just a myth that because a chemical plant is just one huge mass of steel, if a storm goes overhead there will be a lightning strike. The lightning system is there just in case. Saying that, I have known lightning hit a column and ignite the vent.

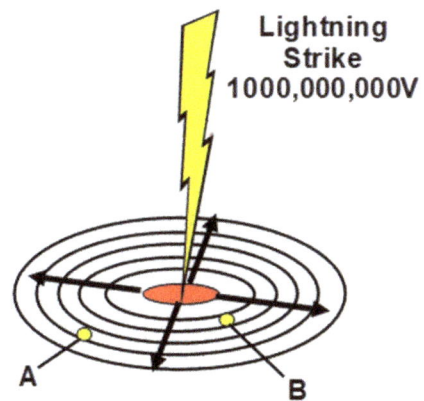

Another phenomenon that happens with lightning strikes is what we will call the dartboard effect. If you look at the diagram on the left and the lightning strikes bull's eye, the potential starts to eke out in the direction of the arrows. There could be a potential difference of many thousands of volts between point 'A' and point 'B'. Just imagine cattle standing on this dartboard with front legs on 'A' and hind legs on 'B', or footballers running on a football field. As they each have less resistance than the ground, many thousands of volts could flow through them. Lightning does not have to directly strike you to kill you.

Lightning conductors do have a catchment area, like an umbrella, meaning that shorter buildings below the conductor may not require lightning protection. See the Standard.

Lightning Protection Zones:

Have you ever wondered why lightning is not straight when you see it? This is because it is electricity and hence follows the path of least resistance in the air. There are moisture clouds and clouds with impurities in the sky which may conduct the lightning more easily than going through pure air which is why it zigzags across the sky.

It will not zigzag once it hits a lightning conductor though. The lightning path will be straight down to ground so it is of no use to follow the contours of a building with your lightning conductors.

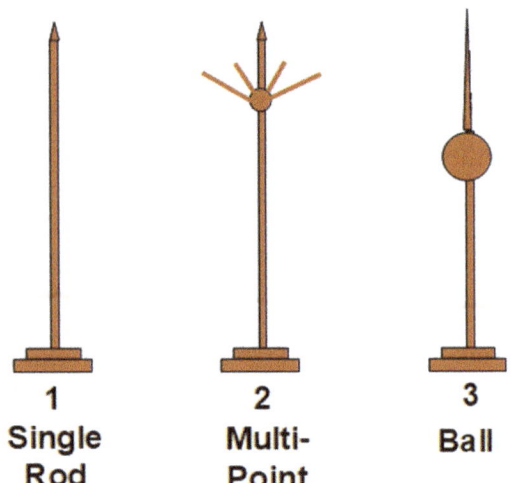

1 Single Rod **2 Multi-Point** **3 Ball**

There are many, many different shapes of lightning rod that you can choose from. I have put three to the left that you may find on your factory.

Usually on the tops of structures it will be the first model, No.1, which is a single rod. The end of the rod does not necessarily have to be pointed, but this shape might direct the charge much better than a round rod or ball at the end which may spread it.

There do not seem to be any hard and fast rules on the shape.

On the last page we talked about the **'Dartboard'** effect of lightning when it strikes the ground. Well now we will talk about the **'Umbrella'** effect.

The diagram on the right shows what I have called the 'Umbrella' Effect where we have a lightning pole that is around 50 feet high and the 'Protection Zone' below.

The buildings or tanks below will not have lightning conductors as they are in the Zone. These could be structures of a plant, or tanks on a Tank Farm. The higher the mast the bigger the Zone.

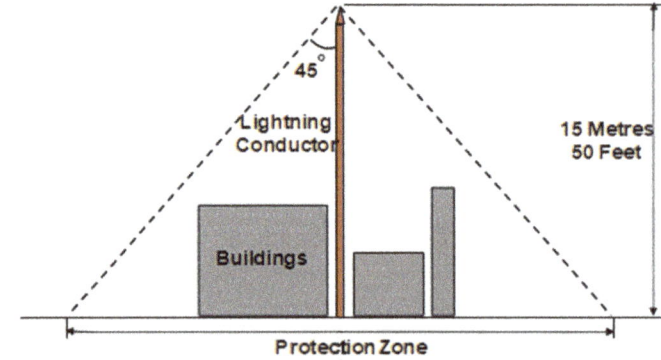

A 100 foot tower would, for instance, protect a circular area of a nearly 200 foot diameter below it.

You would have to look at where the lightning conductors should be on the highest buildings on the plant and work out the Protection Zone if I placed the conductors 'X' number of feet apart. I say the highest buildings as that would protect all of the buildings within the 'Umbrellas'. The lightning conductor does not have to be in contact with the structure. If we take a chemical storage tank, the lightning conductor could be on a pole a certain distance away so long as the structure falls within the umbrella. The next page explains a system that may be installed on a plant.

Risk assessments on surge protects should be completed on vulnerable systems.

Lightning Ring Earthing Conductors:

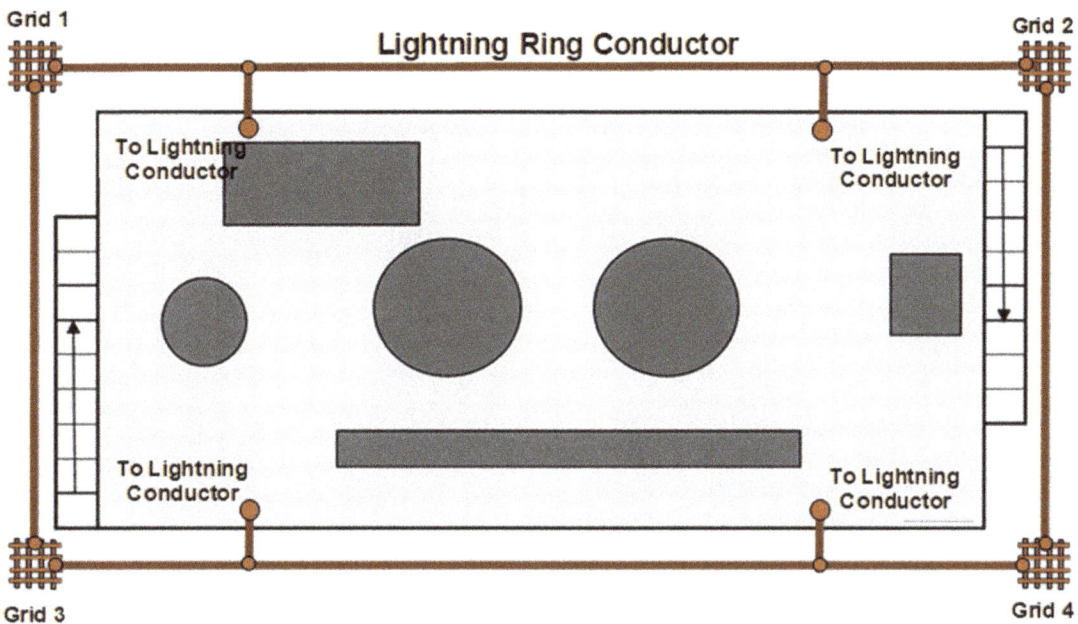

A lightning ring earthing conductor is exactly what it says. A ring copper conductor is run around the structure, possibly with grid conductors at certain points. Rod electrodes can be used and the Standards must be consulted as to distances etc. Earth Pits are provided at each mesh electrode so testing the ring can be completed.

The bar/tape of the copper electrode must be suitable for installing below ground level and corrosion preventer is advisable. The ring must be around 80% in contact with the ground. These are usually used in terrain where it might be difficult to hammer rods into the ground hence the grids that I have put in the diagram above. The ring electrode depth should be around a metre and buried around one metre away from the building.

The ring is meant to disrupt what is called the electromagnetic induction field produced by the lightning voltage/current. The electromagnetic induction is caused by the voltage moving down a stationary conductor; the opposite of a generator. The use of concrete enclosed earth rings must be looked into very carefully because the lightning voltage could cause the damp concrete to explode.

Advice must be obtained before connecting the ring electrode/earth bars to any other earthing protection system such as equipotential bonding. Other services must not be run in parallel with the down conductor, and also distances of other earth electrodes in the ground for other services such as **Intrinsic Safety** should be checked.

The ring, electrodes, conductor and down tapes must be carefully inspected after a known strike. It is likely that a direct strike may cause some devastation of the system, but remember lightning protection systems are there as a prevention measure to try and stop the strike in the first place by releasing positive 'ions' into the storm cloud. Risk assessments on step voltage must be looked into where someone may be unlucky enough to be in the locality during a strike.

1) What form does lightning protection take on your plant?

2) Have you got lightning conductors on tall buildings?

Voltage Surge Protection Devices (SPDs):

The first question that must be asked is 'Does a device that protects against transient higher voltage or lightning surges warrant being installed in the first place? i.e. how reliable is the supply?' The second question must be 'What are we protecting?' Well electronic equipment may be our first port of call as well as computers. Surge Protection comes in many types, but I will explain two main different forms. These can be in the design of what is known as 'Spark Gap Technology' which is very efficient, fast and can handle large voltage such as lightning. The other form which might be better known is the 'Varistor' type which diverts the surge voltage away. Surge Protection Devices (SPDs) are to **Standards EN61643–EN62305.**

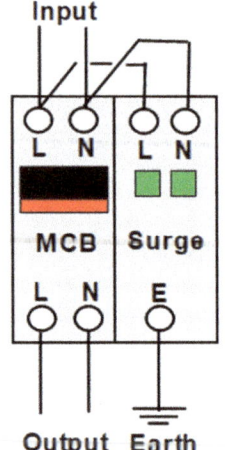

Spark Gap Technology as per the diagram on the left is connected as a 'shunt' and provides a path to earth should the voltage rise above a certain level. Can be used in domestic circumstances i.e. circuit breakers in consumer units. These are sacrificial and the green indicators in the diagram on the left turn red if the module has operated its protection system.

These units are cheap and can be obtained to guard just a single pole or, a bit more expensive, a double pole miniature circuit breaker. They fit easily onto the din rail and so long as there is room for them they are ideal surge protectors.

Let us now look at what is called a **'Varistor'**. (These may well be called Metal Oxide SPDs as per diagram on the right.) Well let us start by saying that the word 'Varistor' comes from two words '**Var**iable **Res**istor'. I can say that a Varistor bends Ohm's Law to a large degree and, unlike a resistor, does not operate in a linear direction. It consists of a sealed outer case containing two electrodes either side of a chamber made up of ceramic with Zinc Oxide pellet shapes.

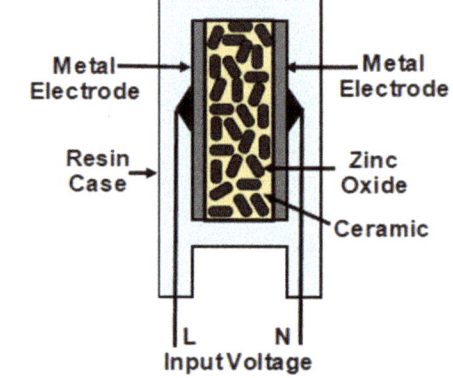

Metrosil Technology uses a very similar process and protects delicate instruments.

A bit like a Zener diode in a barrier, the Varistor will not conduct voltage until it rises to what is called the breakdown voltage. As the voltage increases the resistance decreases exponentially.

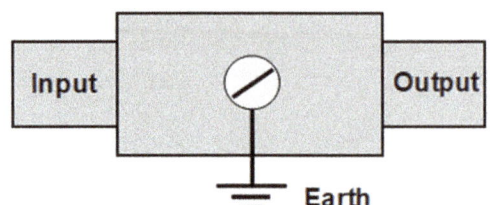

Sometimes surge suppression is required on Intrinsically Safe Systems which can consist of a Metal Oxide SPD as described above or maybe what is called a 'Gas Discharge Tube Arrestor' as in the diagram on the left. With Intrinsic Safety this would be known as 'Simple' Apparatus'.

The tube basically is a gas filled ceramic container with two electrodes inside. Protection depends upon the distance apart of the electrodes and the gas which could be something like Deuterium or Hydrogen. As the gas molecules ionise they become a conductor and provide a path to earth.

1) Have you got Lightning/Surge Protection anywhere?

High Voltage Earthing:

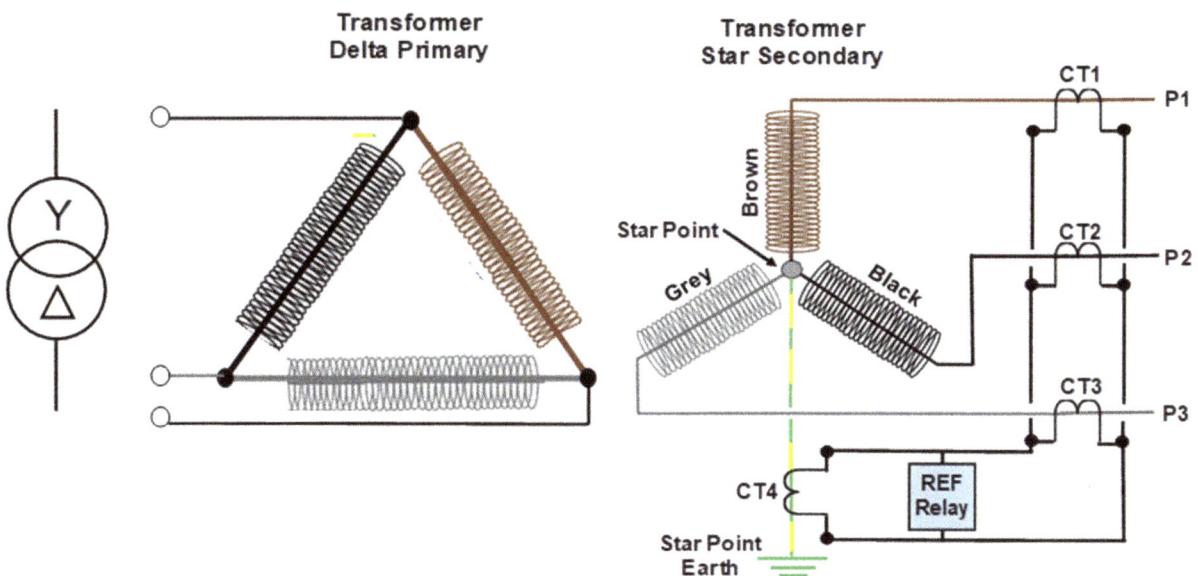

Many of the systems in your Hazardous Area may be high voltage (HV) i.e. Transformers, Motors etc. which must also be protected from earth faults. Also, maintenance has to be completed on this equipment and certain earthing measures must be carried out for personnel and plant safety.

One thing we must protect against is what is called **'Step' voltage.** If someone is standing next to the main earthing electrode of a HV system and there is a massive high voltage earth fault, then just as lightning can cause a problem with ground potential, so can very high voltages. It is possible in an earth fault for two areas of ground around the earthing electrode to be hundreds of volts different in potential. Stepping from one potential zone into another could prove fatal. **Step voltage** is estimated by taking a person with a stride of **1 Metre** and it is what the voltage would be from foot to foot.

We must also be made aware of large earth faults by using relays which send alarms to a central control room. If for instance we only earth the HV panel at one point then we can measure any earth fault currents by putting a CT around the earth output and taking this to an earth fault relay. The relay would trip the incoming circuit breakers at the same time as sending the signal, thus eliminating the step voltage.

If you work on high voltage busbars or cables there are special measures and protocols which must be observed to ensure your safety. I have included a short HV section showing important earthing protection and procedures that you may come across both on site and in your substations.

1) What are the High Voltages you have? i.e. 33KV, 11KV, 6.6KV, 3.3KV

2) Have 'Step' Voltages etc. been worked out?

3) Do you have Restricted Earth Fault Relays?

4) If the answer to '3' is 'YES' where are Relays located?

5) If the answer to '3' is 'YES' where are CTs located?

6) Do you have Bus Zone Protection?

Single Earth Panel Earthing:

As a young Electrical Technician in the early 70s I was on a commissioning team inspecting and commissioning a new 33KV Feeder Substation for my company. The substation was shared between my company and Yorkshire Electricity Board as it was then, who was the electricity provider.

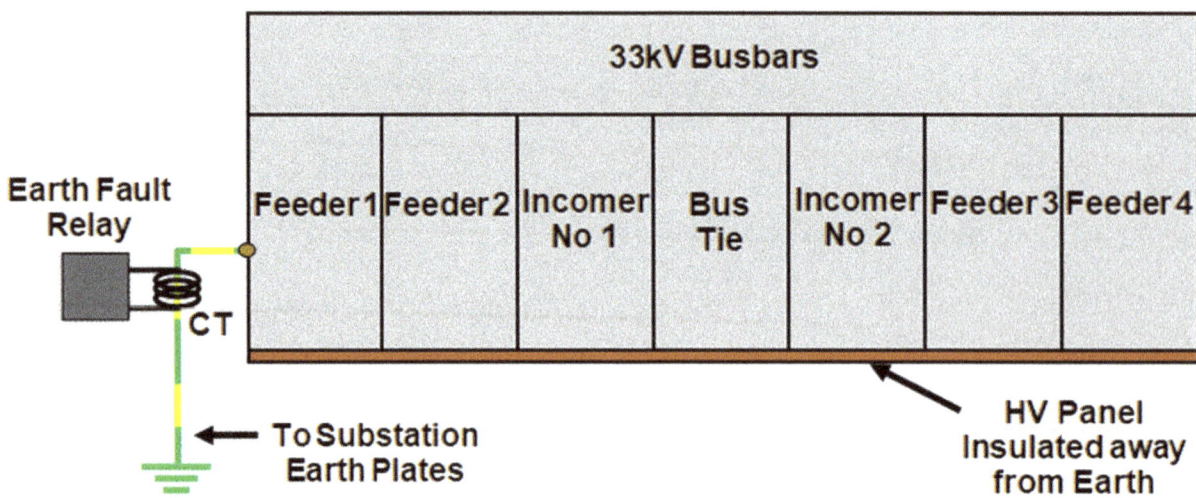

Let me just explain what is happening in the diagram above. The HV panel has 2 x 33,000V Incomers and 2 feeders to one 33KV/11KV substation and 2 feeders to one 33KV/6.6KV substation.

There was huge concern over earth faults at that voltage and whether personnel could be injured by ground or touch potential (Step Voltage). The layout was that the main breaker panel, above, was insulated away from earth and one solitary panel earth busbar was taken to the substation earthing grid made up of earth plates and a lattice. On this earth busbar was a CT operating a relay which would trip the circuit breakers in the event of a fault.

If they just installed a HV panel say 33KV or above and earthed the panel directly down to the rods or earth plates, the danger is that if there was to be a massive earth fault, it is possible for the ground potential to rise considerably and hence it would be similar to lightning hitting the ground and forming different zones of potential. Anyone who stood with their legs apart in two different zones of potential could end up with hundreds of volts going through them depending upon the ground resistance. This may be called 'step' voltage.

By insulating the panel away from earth in this way and taking the earth via a busbar down to the earthing system it must be ensured that nothing was clipped to the panel with an alternative earth wire such as a **gripper hand lamp** etc. Personnel could also be unlucky if they were working on the panel and were in contact with it at the exact moment that a huge earth fault developed. Care even had to be risk assessed if for any reason the panel had to be drilled as the drills of that year were not double insulated.

1) Do you have this type of HV Panel Earthing system on your plant?

2) If the answer above is 'YES' what voltage is the HV?

Bus-Zone Protection:

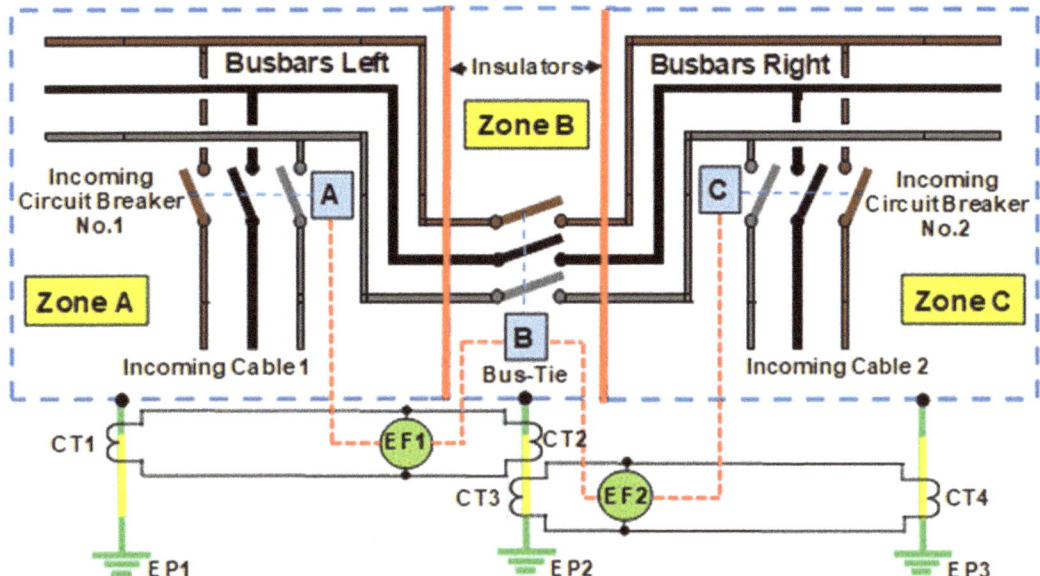

Let me just explain the above diagram which looks complex, but I can assure you it is not. Firstly what we are looking at in this instance is a High Voltage Panel, denoted by the blue dotted lines. This panel has two Incomer circuit breakers **(Blue A and C)** and a Bus Tie Circuit Breaker **(Blue B)** joining up the left hand panel busbars to the right hand panel busbars. The idea of the bus tie is that if for some reason we lost or had to isolate an incoming circuit breaker then the two panel section busbars would stay energised. Or if for some reason we wanted to isolate one half of the panel we could trip the incomer on that side and open the bus tie.

NOTHING MUST BE CLIPPED ON THE PANEL SUCH AS GRIPPER HAND LAMPS THAT ARE EARTHED.

The panel is physically separated into three Zones **(A, B and C)** by two red insulators. These Zones are totally separate from each other and each of the three Zones would have its own earthing point **(EP1, EP2 and EP3)**. Each of these Zone Earths has a Current Transformer monitoring any Earth Currents **(CT1, CT2, CT3 and CT4)**. These CTs are connected in pairs to the central Earth and in parallel with an Earth Relay **(EF1** for the left hand pair and **EF2** for the right hand pair).

If for instance we had an earth fault on the left hand side of the panel **(Zone A)** we would want to remove all power from that side of the panel so we would want to open **Incomer No.1 Circuit Breaker 'A'** and the **Bus Tie 'B'**. If we had an earth fault on the right hand side of the panel **(Zone C)** we would want to remove all power from that side of the panel so we would want to open **Incomer No.2 Circuit Breaker 'C'** and the **Bus Tie 'B'**. If we had an earth fault in the mid-section **Zone B** we would want to trip both incomers. Earth Fault Relays **(EF1 and EF2)** are wired in parallel to the CTs **(CT1 and 2)** and **(CT3 and 4)** and the contacts of the Earth Fault Relays are connected to the trip circuit of the three circuit breakers. Now any earth fault on one of the three Zones **(A, B and C)** will trip the appropriate circuit breakers and isolate the Zone quickly before the fault, hopefully before it has chance to do much damage.

1) Have you got this type of HV Earthing Protection on your plant?

Restricted Earth Fault:

When we look at high voltage, usually we are also looking at high current, and faults to earth can cause the current to go exceptionally high so we must have protection systems that are extremely efficient and fast to prevent damage.

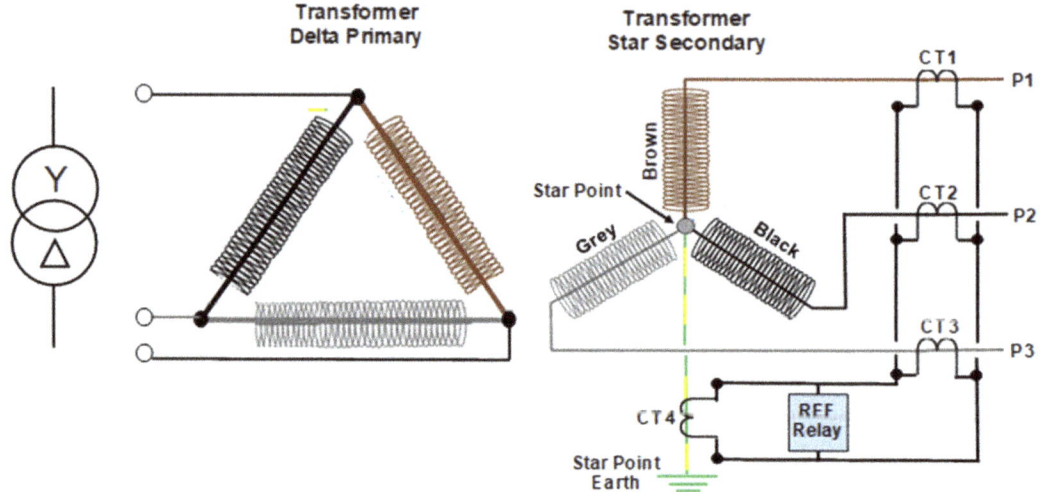

Many transformers on your factory will have **Restricted Earth Fault Protection** so what is this? Well the word **'Restricted'** may be a clue; the fault is restricted to a certain part or 'Zone' of a circuit. It is all a matter of balance. We put a current transformer around each of the phases, **CT1**, **CT2** and **CT3** above, then we put another current transformer around the earth on the star point of the transformer secondary winding **CT4**. All earth faults flow from the phase through the ground and up the star point earth through **CT4**. These current transformers are connected together in parallel as in the above diagram to a **Restricted Earth Fault Relay**. So if I were to give an analogy, it would be a bit like an earth leakage protection relay where, so long as there was no earth fault, the system would be balanced so the 'Vector Sums' of all of the CTs would be **ZERO**.

If however an earth fault appeared on one of the phases within the zone from the CT to the transformer, then that CT would become out of balance with the rest of them resulting in current flow through the Restricted Earth Fault Relay which would trip the system before significant damage was done. Hence as I said earlier we could 'Restrict' the Earth Fault. Tripping the system would of course mean that the circuit breaker feeding the transformer would trip and inter-trip the circuit breaker downstream.

At our BP factory the restricted Earth Fault system was mainly on the delta star distribution transformers and the CTs were located in the LV distribution panel in the switch room. Locating them here increases the system area that they protect. Too close to the transformer limits the protected area or zone to a very small range, even maybe just the windings themselves. The sensitivity of the protection within the zone must be extremely low, maybe less than 20% of the normal current flow, to avoid the damage that an earth fault could cause to the system. The **Earth Fault Relay** has a **'Stabilising Resistor'** fitted to stop earth faults outside of the zone of protection from inadvertently tripping the relay.

1) Do you have these relays on your plant?

2) If 'YES' are you aware of where they are located?

Earthing Transformer & Delta Earth Protection:

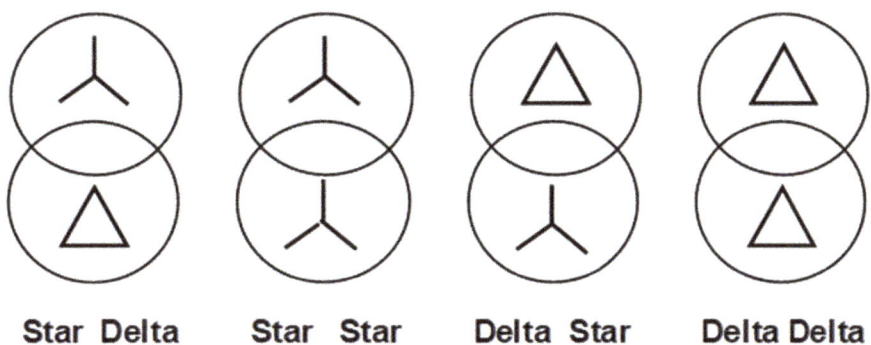

HV transformers can be configured in four different ways, as shown above, for a whole host of reasons. It may be that, in the positive reasons, harmonics are reduced, less transformer windings in star etc.

The secondary is star so that it is easy to earth the star point and form a safe neutral in TNC, TNC-S, TNS, TT & IT Systems etc.

The negative reasons may be that delta secondary windings require more wiring, there is more chance of a phase shift that may have to be rectified, and protection from earth faults can be difficult without going to a lot of expense in equipment, as below.

Star windings are a simple way to provide an earthed neutral, but what if the secondary winding is delta? There may be a question to be asked: 'If this is a distribution transformer, how do we earth the system safely?'

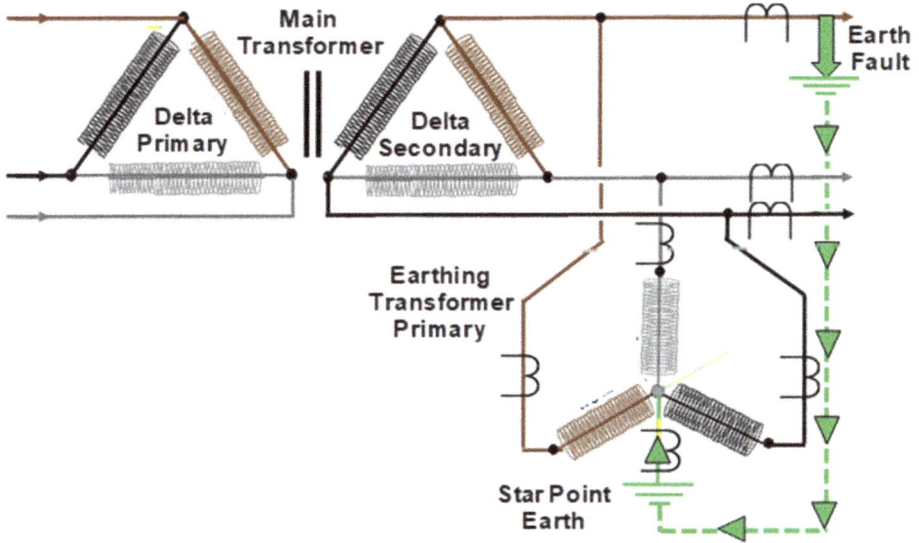

One of the ways we can earth the delta secondary safely is to insert another transformer which does not do much except provide a star winding with its star point earthed. In this way we provide a path for the earth current, and with strategically placed CTs we can detect this earth fault and use a relay such as a restricted earth fault relay to trip the system and stop the earth fault from doing a lot of damage.

Providing another large transformer to obtain the neutral and star point can be a very expensive way of doing things so 'star' secondary windings are much preferred.

1) What configuration are the HV Transformers on your plant?

High Voltage Busbar Earth:

Now and again Routine Maintenance requires that work is done on the busbars. It might be checking tightness or just an inspection of the condition of the copper.

To do this safely, certain procedures have obviously to be carried out to ensure personnel safety from high voltage accidents. **The Procedure below is a suggested example and not meant to go against any Company Policies.**

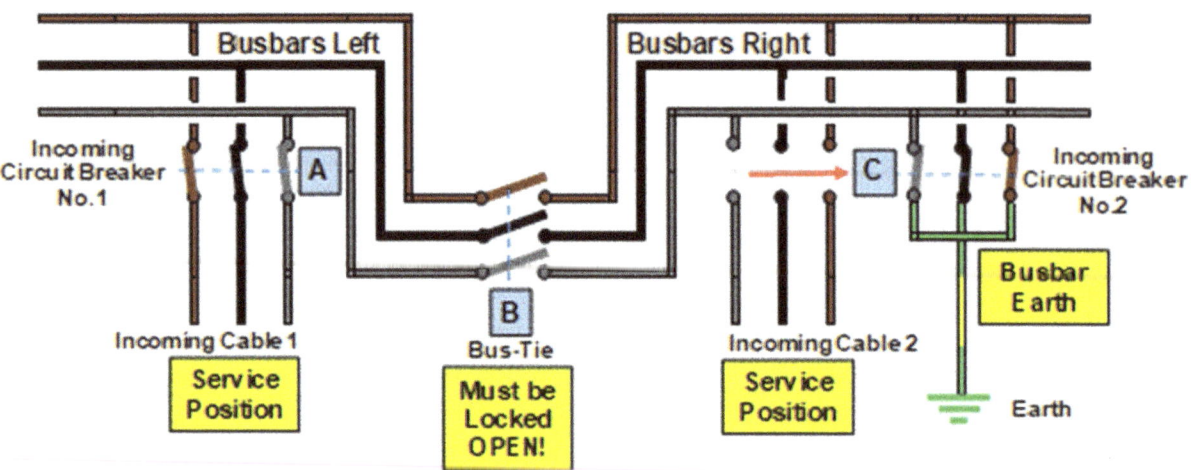

Above there is a diagram of an HV panel; let us just say that this panel voltage is 3.3KV. The job in hand is that as the Electrical Engineer I want a Team of Electrical Technicians to check the tightness of the bolts which bolt the busbars together on the right hand side of the panel.

I have completed an Electrical Permit, carried out a Risk Assessment and completed a Method Statement (Safe System of Work) so all documentation has been read and signed by the Electrical Technician Team carrying out the work and the Electrical Team Leader in charge of the Team and we are ready to complete the isolation.

Team is briefed on the work by the Electrical Engineer and Team Leader. Any special requirements i.e. torque settings should be declared.

For this exercise we will require something called a **'Lock-out Box'**. This is a simple box with two hasps on for two locks. The Electrical Engineer has one lock and the Electrical Technician Team Leader has the other lock. In this box go all of the lock on/lock off keys that fit the locks on various circuit breakers to make the system safe.

Certain circuit breakers such as the bus-tie would have to be locked in the open position so that it cannot be closed, and things like busbar earth would be locked in the closed position and, as noted above, the keys would go into the lock-out box. The Electrical Engineer and Team Leader each fit one lock on the lock-out box. Without **both** people being present the keys cannot be obtained to operate the circuit breakers.

1) Do you carry out work on High Voltage Busbars?

2) If the answer to the above is 'YES' are the busbars earthed?

3) Have you ever seen a Lock-out Box?

High Voltage Circuit Earth:

Now and again Routine Maintenance requires that work is done on the Cables or Transformers. To do this safely, certain procedures have obviously to be carried out to ensure personnel safety from high voltage accidents. **The Procedure below is a suggested example and not meant to go against any Company Policies.**

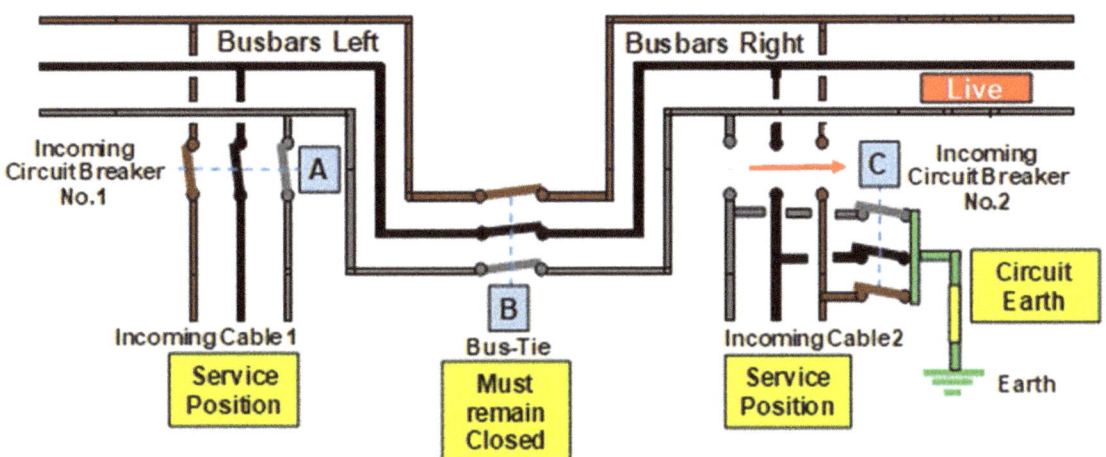

The moving portion of the circuit breaker is tripped, unplugged and wheeled to the **'Circuit Earth Position'**. Above there is a diagram of an HV Panel; let us just say that this panel voltage is 3.3KV. The job in hand is that as the Electrical Engineer I want a Team of Electrical Technicians to check the transformer and/or feeder cable of incoming circuit breaker No.2. In this case the **Bus Tie Circuit Breaker** will remain CLOSED so that it feeds the right hand side of the panel as we have removed the **Incoming Circuit Breaker**.

I have completed an Electrical Permit, carried out a Risk Assessment and completed a Method Statement (Safe System of Work). So all documentation has been read and signed by the Electrical Technician Team carrying out the work and the Electrical Team Leader in charge of the Team, and we are ready to complete the isolation.

Team is briefed on the work by the Electrical Engineer and Team Leader. Any special requirements i.e. Test Equipment to be used, spare parts to be fitted should be declared

For this exercise we will require something called a **'Lock-out Box'**. This is a simple box with two hasps on for two locks. The Electrical Engineer has one lock and the Electrical Technician Team Leader has the other lock. All lock off keys for the whole job, Circuit Breaker Positions, trip Mechanisms, Bus Shutters etc. are put into this box.

The Electrical Engineer in charge of the project will put their lock in one hasp and the Team Leader in charge of the Team of Technicians will put their lock in the other. Without **both** people being present no de-isolation can be completed. With the Circuit Earth, as mentioned above, the work may be on a transformer or a cable joining substations in a circuit in which case there will be a circuit breaker on the other end that will require **Circuit Earth** etc. Maintenance is usually carried out on both circuit and circuit breakers to save having the system off again in the future.

1) Do Engineers perform this procedure for work on High Voltage Cables?

2) Have you ever worked on a High Voltage Cable?

Transformer Earthing:

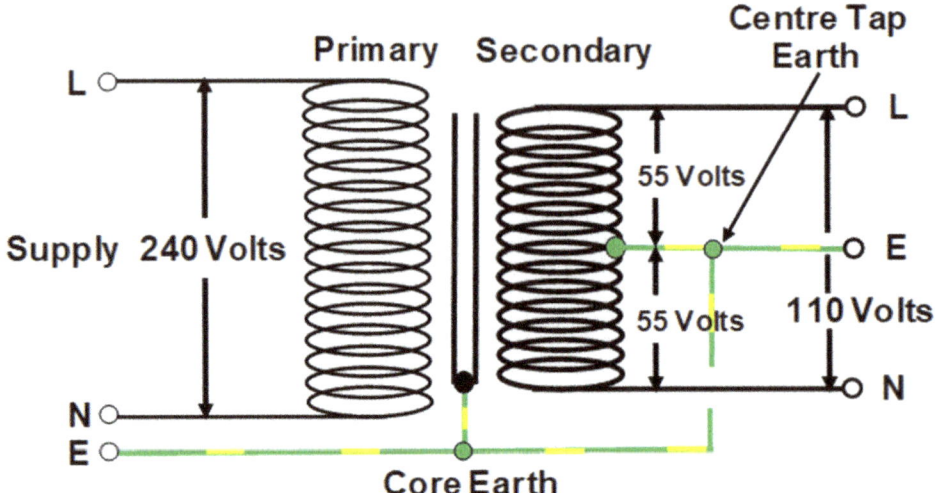

Transducers are devices that take one medium such as electricity and use it to control another medium such as air. A transformer is an **AC ONLY** device that takes one voltage and transforms it into another voltage, either stepping voltage down or stepping voltage up.

Transformers usually have a primary winding and a secondary winding. AC is put onto the primary winding at 50Hz, meaning the voltage changes direction at the rate of 50 cycles per second. As the primary winding magnetic field is set up with its lines for flux, these lines cut through the secondary winding oscillating at 50 cycles per second and induce a voltage into the secondary winding (called an EMF).

If we just had two coils next to each other the transformer would not work very efficiently, if at all, so we install the windings wrapped around an iron core. Just as I explained in my book 'Motors in Hazardous Areas', we laminate the iron core to stop eddy currents and hysteresis causing heat. So back to our windings, to step down the voltage there will be fewer turns on the secondary than the primary, but they will be heavier because of more current.

Sometimes the windings are wound one on top of the other with a metal separator between them. This metal separator is earthed so that if the primary winding was to melt then the primary voltage could not melt onto the secondary without going through earth first and thus the primary voltage could not be put onto the output.

We have discussed earthing systems and the earthing of large transformers such as a distribution transformer where the star point is earthed directly or through a high impedance. There is also a whole range of different smaller transformers shown in this section with many different earthing systems designed to keep us safe.

1) How many types of transformer have you on your plant?

2) Are they earthed correctly after reading this section?

3) Do you wire 110V sockets on the back of your instrument supplies panels?

4) Do you use step up transformers with no ELU/RCD protection?

5) Have you got Instrument Transformers with Floating Supplies?

Distribution Transformer Earthing:

Firstly let us have a look at a typical Distribution Transformer which supplies your Hazardous Areas via substations (HV) and switch rooms (MV). You will notice that there are 3 phases, a neutral and an earth. This transformer is usually connected in what is known as 'Star' and has a point where one end of the three windings are connected together called, as you would imagine, a 'Star Point' where we also get our neutral from.

For safety reasons this 'Star Point' is earthed to the ground with an earth reading of **1Ω** or better and this earth forms the return path for all faulty equipment or humans through the ground and back to the distribution transformer.

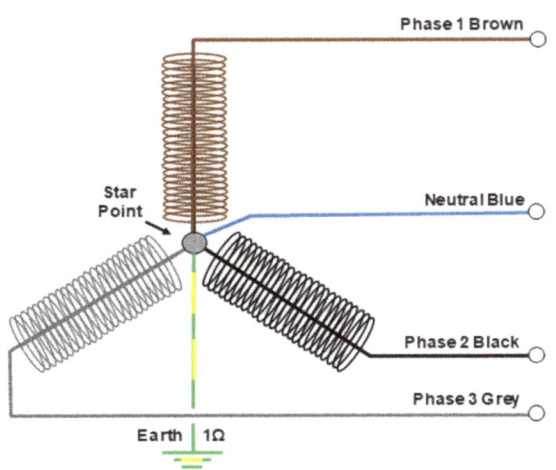

Looking at the diagram on the left we see the 3 phases which these days are brown, black and grey (replacing red, yellow and blue).

The voltage between each of these phases is around **415V**, however, some years ago this was **440V**.

The voltage between any of the 3 phases and neutral is around **230V**. Some years ago this would have been **250V**.

The voltage between any of the 3 phases and earth is around **230V**. Some years ago this would have been **250V**.

One question that is regularly asked is that; if the safety earth is 230V between it and a phase, the same as the neutral, and is the cause of many electrocutions, why not remove it? The answer is we can, and almost do, with the **'IT'** Earthing System which I will explain later. The only problem with removing it is that many protection systems that guard faulty equipment rely on that earth reference being present.

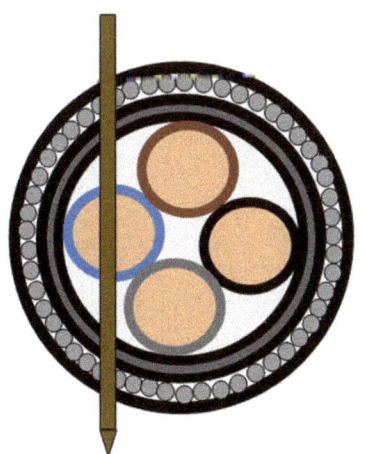

Let us just imagine that the star point earth is removed on the distribution transformer. You might say now that I will never get another potentially fatal shock to earth and you would be correct, **PROVIDING** that the transformer neutral which also emanates from the star point **NEVER** got a fault to earth. If this were to happen then suddenly there is an earth reference and a 230V potential from live to earth. Because you were not expecting this you may have no metal connected to earth designed into the circuit, making faults to the case of equipment lethal. This could happen if, say, the telephone company hammered a spike into the ground without scanning and penetrated the cable and went through the neutral without the other phases.

I appreciate that this might be picked up on some protection relays such as a Solkor Relay, but they usually come into play if the spike was to go through the phase conductor.

Centre Tap Transformer Earthing:

These transformers are centre tapped for safety. Sometimes, especially behind an instrument panel, there may be an urge to install some sockets for test instruments. Be very careful that the 110V used is centre tapped as 110V to earth could be lethal.

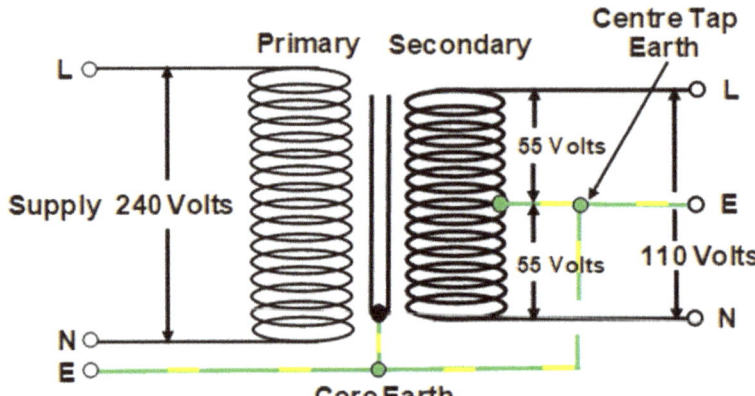

Above I have shown a diagram of a 110V transformer, the common yellow type used for portable tools such as electric drills and grinders etc. You will notice that this transformer has two earths; one earth, as most transformers, have is to earth the transformer core and the other is to the centre of the secondary winding which is called a centre tap earth. Just a point; using 110V tools on this transformer is unlikely to cause fatal injury.

Let us first look at the transformer core. This is earthed in case of a fault in the windings of the transformer. If the primary winding was to melt then the last thing we would want is the primary voltage winding melting into the secondary winding giving us a primary voltage on the output. Here, if the winding was to melt, it would have to go through the earthed core to get to the secondary, hence blowing the fuse first.

Secondly we have the centre tap earth on the secondary winding. This ensures that, in the case of someone coming into contact with the output voltage, they could only get a shock of 55V to earth. Although on the secondary I have put 'L' (Live) and 'N' (Neutral), in actual fact we just have two legs of a transformer, but it just gives the technician who is connecting up the sockets a reference for Live and Neutral to correspond to the 'L' and 'N' in the socket outlet. Although I have shown a 'tool' transformer in the diagram with a 55–0–55V output, centre tap transformers have many other uses besides tools and can be obtained, for instance, with a 110–0–110V or a 12–0–12V output.

The common yellow transformer used for tools on factories and building sites has a very sturdy case. It is usually ingress-protected to IP44 (Particles Greater than 1mm and Water splashed from any direction) so the transformer is ok in the rain. The socket has a spring lid that springs closed and this also stops the plug being accidently pulled out by having a catch that fits the receiver on the plug. These lids get broken off and should be repaired immediately as the ingress protection is affected and the plug could arc if pulled out. There is also a thermal overload, **NO FUSES**, in case the transformer current goes too high. The output sockets can be 16A or 32A depending upon the transformer size. Checks should still be made for the centre tap.

1) Are all of your Yellow Tool Transformers centre tapped?

Instrument Supply Transformer Earthing:

Sometimes, behind an instrument panel, there may be a tendency to put some sockets onto the 110V for test instruments.

THIS MAY NOT BE A CENTRE TAPPED SUPPLY FOR TOOLS!

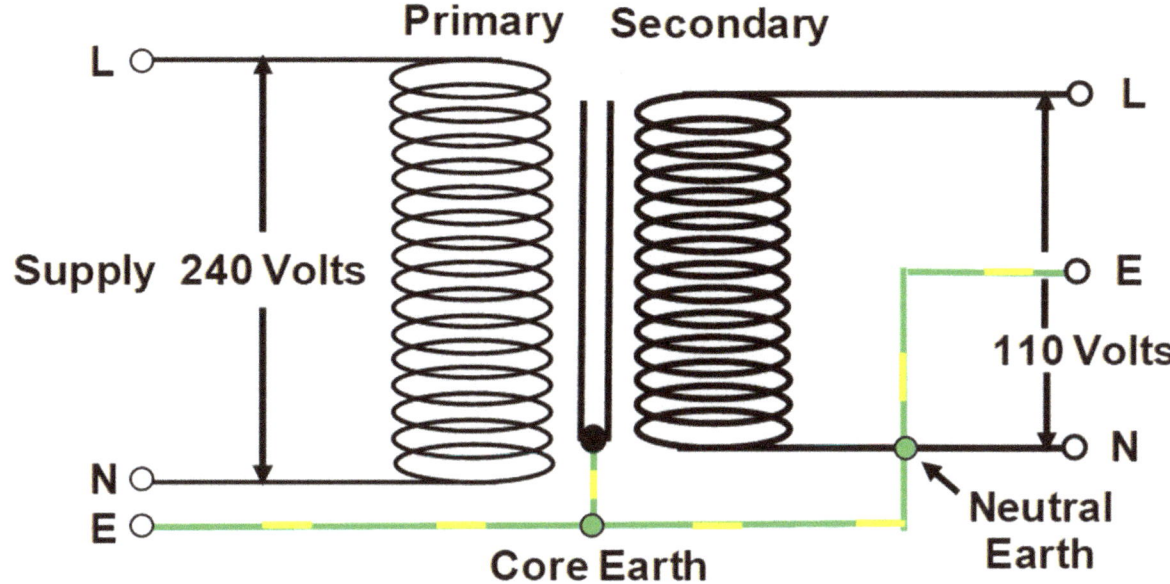

Above I have shown a diagram of a 110V transformer of the type used for Instrument supplies etc. You will notice that this transformer has two earths. One earth, as most transformers have, is to earth the transformer core and the other is the neutral of the secondary winding.

This transformer could be any voltage but I have kept it at 110V to show you how very different the secondary is from the centre tapped transformer we looked at earlier.

Let us first look at the transformer core. This is earthed in case of a fault in the windings of the transformer. If the primary winding was to melt then the last thing we would want is the primary voltage winding melting into the secondary winding giving us a primary voltage on the output. Here, if the winding was to melt it would have to go through the earthed core to get to the secondary, hence blowing the fuse first.

Secondly we have the secondary neutral earth. This ensures that the system has an earth reference in the case of a fault, but in this case there would also be 110V to earth which is a lethal voltage so **no power tools should be used on this supply**.

As mentioned above, this type of transformer could really be any voltage 240/415 on the primary and any lower voltage on the secondary. Again, this secondary voltage would be the same for live to earth as it is for live to neutral. This may not always be possible depending upon the transformer, but a floating supply can be obtained here by removing the earth from the neutral on the secondary.

1) Have you got sockets connected into your 110V Instrument supplies?

2) If 'YES' I assume that Tools may be plugged in?

Floating Transformer Earthing:

In a standard electrical supply, because of the star point earth on the distribution system, on a TT or TN-S earthing system there is 240V between live and neutral and 240V between live and earth, but what if we had a supply where there was no star point earth so there was 240V between live and neutral, but not live to earth? We are getting very close to this with the IT earthing system. Smaller supplies and transformers require floating supplies for a whole range of reasons, which might be, for instance, removing electrical noise from the circuit for various instruments. These are usually also on Class III Equipment (PAT Testing). Class III equipment really does not require PAT testing at all, just the leads. (Check Manufacturer Advice.)

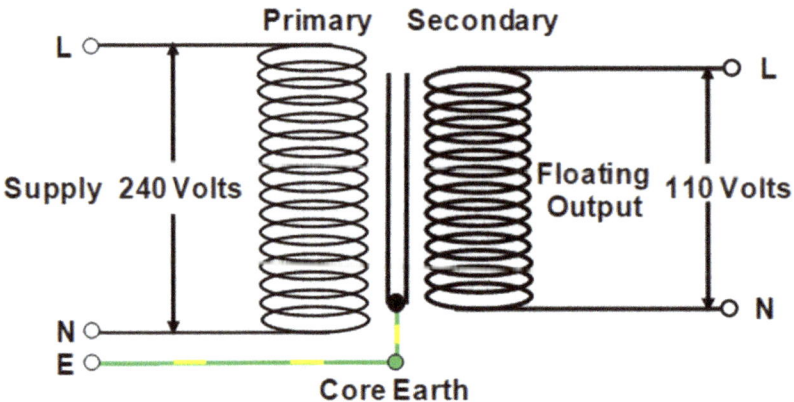

Above I have shown a diagram of a 110V transformer with what is called a 'Floating' output where the neutral of the output has no reference to earth. This may be called an 'Isolation' transformer.

Let us first look at the transformer core. This is earthed in case of a fault in the windings of the transformer. If the primary winding was to melt then the last thing we would want is the primary voltage winding melting into the secondary winding giving us a primary voltage on the output. Here, if the winding was to melt it would have to go through the earthed core to get to the secondary, hence blowing the fuse first. The 240V supply on the supply side of the transformer will be referenced to earth because of the star point on the distribution transformer.

You may find floating outputs on equipment such as plugs with USB ports on them, which are used for charging things like iPads etc., where there must not be any fear of voltages to earth. If you look at many charging plugs etc. the manufacturers do not put fuses in them so this is just another way to ensure safety. In many electronic recording instrument and test instrument supplies these days the manufacturers will insist on a floating supply and sometimes supply a unit with the instrument to ensure just that.

A good example is a bathroom shaver socket which at one time were quite popular. The output to the shaver is floating so you could (please do not try this at home!) get hold of the live of the shaver socket and get hold of the tap and not get a shock because the output was floating and they just used a 1:1 transformer with no neutral to earth. The secondary supply can, sometimes, be changed to earth referenced by earthing the neutral. THIS MUST ALWAYS BE DONE WITH THE MANUFACTURER OR ELECTRICAL ENGINEER'S APPROVAL!

Step Up Transformer Earthing:

Transformers can of course step up as well as step down. Some companies would not allow these transformers onto their complex.

The company may have a policy where they state that every 240V socket on their site is guarded by earth leakage units. Of course there would be not be one here unless a separate unit was applied.

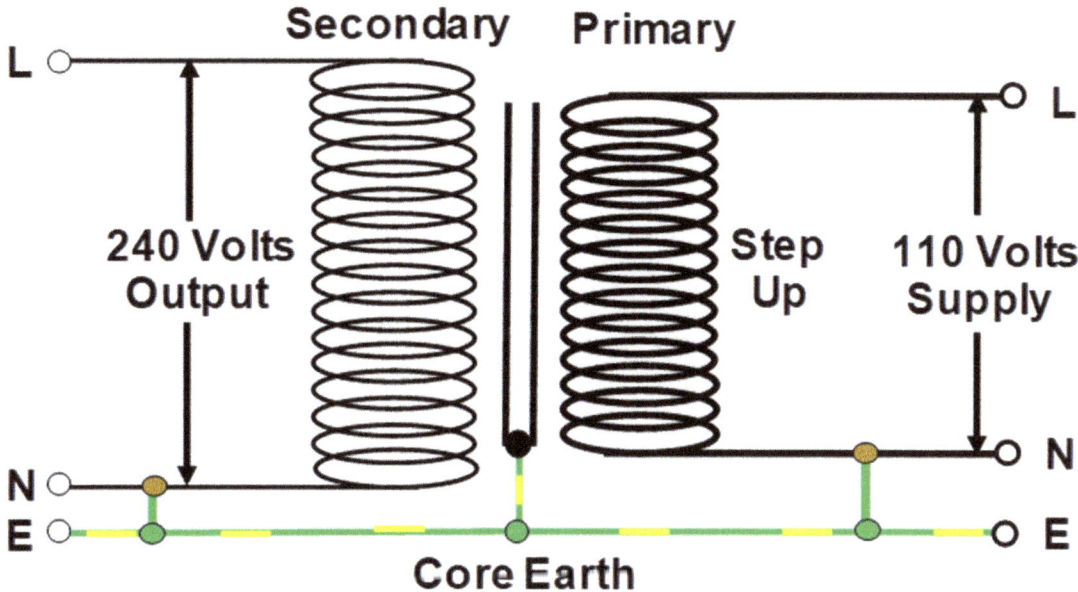

Above I have shown a diagram of what is called a 'Step Up' transformer with a 110V supply stepping up to 240V. Of course transformers can be obtained which step up from and to many different voltages. I have used 110V as an example. These transformers can be obtained in the form of the common yellow block transformers used for 110V tools except in this case the supply plug would be yellow and the output sockets would be blue.

As above, care should be taken in this direction in industry as many companies have a policy that **all 240V sockets are protected by earth leakage or RCD** and in this case they are not, unless a separate unit is plugged into the output.

Let us first look at the transformer core. This is earthed in case of a fault in the windings of the transformer. If the primary winding was to melt then the last thing we would want is the primary voltage winding melting into the secondary winding giving us a primary voltage on the output. Here, if the winding was to melt it would have to go through the earthed core to get to the secondary, hence blowing the fuse first.

The supply in this case will have a lesser number of windings than the output and will be of higher cross-sectional area wire because of the higher current involved. The transformer turns ratio for a step up transformer is more or less proportional to the voltage ratio.

Power stations use step up transformers for the national grid voltage, stepping the voltage up to 400KV from around 200KV before it leaves the power station.

1) Have you authorised step up transformers on your plant?

Auto Transformer Earthing:

These transformers are very handy and cheap, but they must NEVER be used where there would be a dangerous situation if the output volts suddenly, for some reason, shot up to equal the supply volts in the case of a fault.

Let me use the following example, which I may say you are unlikely to have in your Hazardous Area, **a train set**. The input of the auto transformer is 240V and the output 12V. It would obviously be very dangerous if the output suddenly went up to 240V because of a short circuit on the coil.

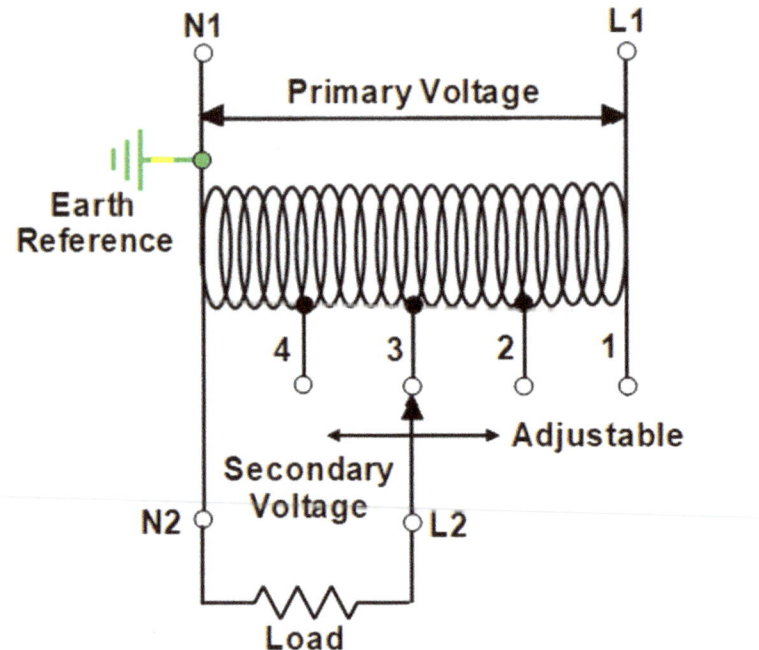

These transformers can be used for tasks like pipeline trace heating control where the output voltage would vary the voltage depending upon how hot the trace heating had to be. If the voltage suddenly went to maximum it would simply make the tracing hotter or put it onto boost.

Most transformers that are mentioned in this book have what is called an earthed core which is a safety device. If any fault occurred on the transformer primary or secondary and the windings were to melt then for the primary winding to melt into the secondary winding and give the full primary output, the melted windings would have to go through the earthed core, hence blowing the fuse.

Well in the case of the Auto Transformer there is no such safety mechanism so if you look at the above diagram, if the winding was to melt from position 1 to position 3 the Auto Transformer would give out **FULL** primary voltage.

There is an Auto Transformer soft-start application for electric motors mentioned in my book 'Motors in Hazardous Areas'. Again there is not much danger with this application if the output went to full primary voltage. All components in the circuit would come to no harm although the fuses may blow. A **Variac** is of course a variable auto transformer.

1) Do you use Auto Transformers?

2) Are you happy with the safety of this Auto Transformer?

Current Transformer Earthing:

I think that, before we go onto the next stage, I should explain what a current transformer (CT) is. The current transformer (CT) is obviously an AC device consisting of a coil where a conductor passes through it as in the diagram below. Unlike a voltage transformer that we all know, this device deals in current.

All I can say about earthing here is: Earth as per manufacturers' Wiring Diagram.

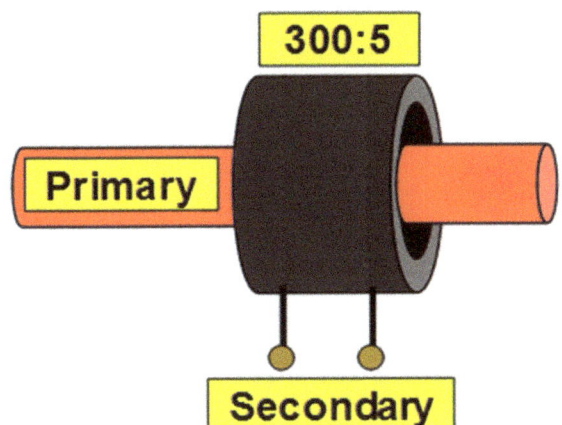

In the diagram on the left you can see a coil, with a laminated core, around a conductor (Red). In transformer terms, the conductor is actually the primary in this case and the windings of the coil the secondary. The idea of the device is to take a high AC current in the primary i.e. 300A and drop that to a low current in the secondary 5A, so in our diagram the CT ratio is 300:5. By doing it this way I can have lower current devices i.e. ammeters or overloads on high current systems.

The current in the secondary (coil) will be proportional to the current flowing in the primary (conductor) so we can actually come up with a ratio of: 300:5. As mentioned above I could have a 0–5A ammeter measuring a 0–300A circuit. I have picked 300 but the value of measured current can be thousands of amps. The secondary is usually 5A so larger units may be 1000:5.

Let us take a current transformer connected to an ammeter or an overload element. This would be what is called a closed loop system or short circuit. The last thing that you would ever want to do with a CT is open circuit the secondary, i.e. take the ammeter away, and leave the CT in circuit with open circuit secondary.

THIS IS DANGEROUS!

If the secondary device is removed, a short circuit should be put on in its place. Do not leave an open circuit.

The result of an open circuit secondary is that a very high voltage is produced in the secondary. What you have actually done, by removing the load and causing an open circuit and leaving the CT connected in, is to take away the impedance of, say, the ammeter, replacing it with an open circuit. What the impedance is now in the open circuit would be classed as infinity so the secondary would be driving current into an infinite resistance.

The result of the above action would be, as mentioned, a very high voltage on the secondary terminals of the CT, possibly many thousands of volts, which could cause serious electric shock harm to anyone working on the system.

Also because the voltage would, in fact, keep on rising it would eventually reach the breakdown voltage of the insulation of the CT itself and one of two things may happen: either the CT would explode causing the terminals to arc, or it would catch fire.

1) Do you realise the danger of an open circuit Current Transformer?

Generator Earthing:

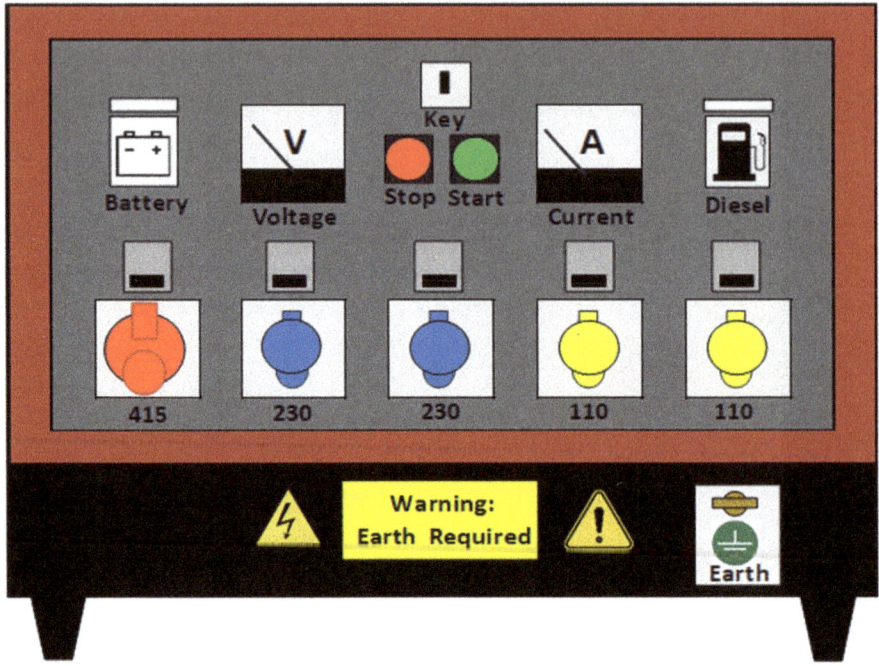

Does a generator require an earth? **Manufacturers will give guidance** here, for instance if I am only using Class II equipment i.e. Double Insulated, then there does not seem much point in an earth. Manufacturer's Guidance in some cases states that if only **ONE** Class I tool is used, an earth is not required. Also see HSE Guidance?

What you have to ask yourself is; how will the protection systems work if you do not earth the unit? Earths are not usually required until something goes wrong. How would the 110V Sockets obtain their centre tap safety without the earth? Can I class the frame of the generator as earth? You may say that they do not require one without a connection to earth, which is all well and good until an accidental earth appears on the system such as a spike through a cable.

Back in the past some companies would not allow welding generators to have earths because, as you will see when you read on, there are certain faults that can burn out a distribution system. All I can say about generators is that to ensure that the safety systems work etc. follow manufacturers' & HSE Guidance as to whether it should be earthed to the plant or have a rod arrangement. However, sometimes factories do not like rods randomly hammered into the ground because of underground services.

It might be, depending upon factory policy, that a wire is taken from the generator earth point to a plant earth bar or rod. I would say that a 4mm cable may suffice which would also be robust, but BS7671 (18th Edition) will give details on cable size.

1) Do you have Portable Generators on your plant?

2) Do you earth Portable Generators?

3) Do you check for the Centre Tap on the yellow 110V Sockets?

4) Are all tools used with this Generator 110V?

5) What protection has the Generator got for 415V & 240V Sockets?

Generator Earthing 1:

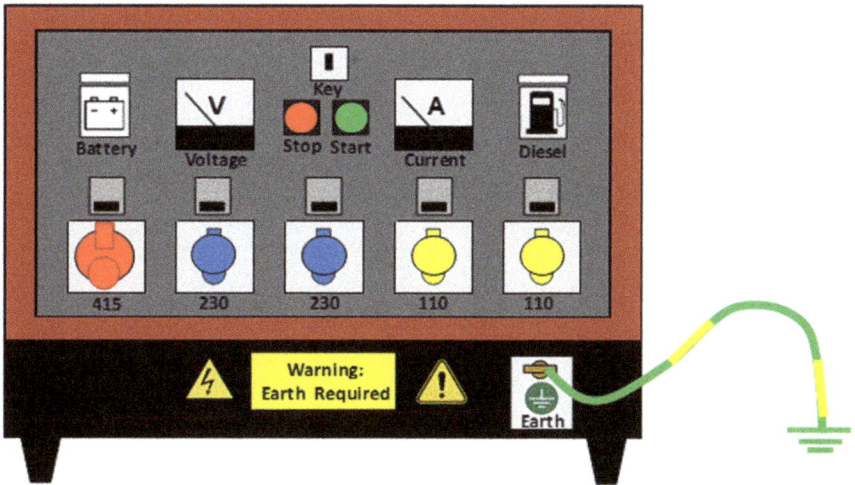

Above I have drawn a diagram of a very simple power generator. As you can see there is a 400–415V Socket Outlet, two 230V Socket Outlets and two 110V Socket Outlets. **Ensure that the yellow 110V sockets are centre tapped.** The generator label or documentation should state whether an earth is required. If a point is provided then it is possibly to earth the neutral. Check the system carefully. TNC NOT in Hazardous Areas!

As the unit stands **WITHOUT AN EARTH**, if I plugged anything into the 240V sockets then I would have a similar arrangement to an **IT Earthing System** with high impedance or no earth on the star point. This might be acceptable if I was just using **ONE** tool or piece of equipment, but would not be suitable if several pieces of equipment were plugged in to the multi sockets.

Without an earth what we have, in actual fact, is what is called a **'Floating Neutral'**, which may on the face of things appear to be acceptable but on a multi equipment basis, where is the earth return path in case of a fault? How, for instance, do the 110V sockets achieve a centre tap to earth when the earth does not exist? Are we using the generator metal frame as the earth? Could problems occur on a floating neutral if a neutral earth fault appeared on one of the output leads and accidentally earthed the neutral?

Some generators, especially American, have a device that plugs into, say, one of the 240V Socket outlets and earths the neutral. This I suppose, if we are not careful, could be a variation of a **TNC** system so may be unsuitable for Hazardous Area temporary supplies.

Check the manufacturer's documentation about earthing our generator either by connecting to a plant earth with, say, an impedance of around **5Ω** or inserting an earth rod in the ground local to the generator. If the documentation says to use a rod, then it should really be tested for impedance. Also if rods are going to be hammered into the ground it would be advisable to carry out a ground scan or whatever company policy states.

Ensure that calculations are completed when relying on RCDs as they may not operate fast enough, if at all, if they are close to the generator with really long leads. This might cause a problem if company policy was to state that all 415V & 240V socket outlets shall be protected by Earth Leakage or RCD. This would mainly apply if you did not earth the generator, thus making a **TN-S Earthing System**.

Generator Earthing 2:

Concerning earthing, a strange fault occurred when I was an apprentice many years ago, when the whole of the distribution system wiring caught fire in a workshop on our chemical factory. Below is an explanation of how that happened.

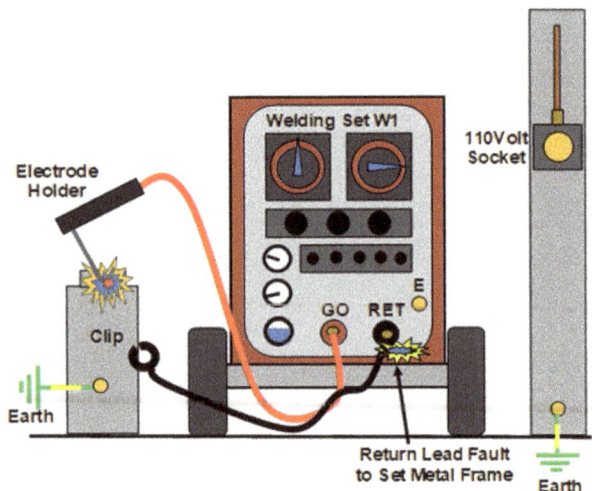

The welding generator, left, was one of the very old type which was on large wheels and had to be towed behind a truck of some kind to the location of the job. The set, obviously, had a 'GO' lead which went to the electrode holder, and a 'RETURN' lead that went to the metal being welded. The set had not been overhauled for some time and had a fault (short circuit) to the frame of the set. This did not cause a problem as the welding current of around **80A** still went down the 'GO' lead and back to the set via the 'RETURN' lead.

Let us now look at a scenario of what can happen when several circumstances come together at the same time i.e. a broken return lead, a fault to the case of the welding set and some metal equipment resting on the top of the set.

The fault occurred when the cable of the 'RETURN' lead broke off the clip, (bottom left on the diagram right) so now there would seem no way back to the generator for the **80A** welding current. A metal drill (Blue) was plugged into the 110V socket and was resting on top of the set. The welding current now had a path back, through the distribution system earthing and down through the socket along the drill flex, through the metal drill to the set case, through the short circuit and back to the generator.

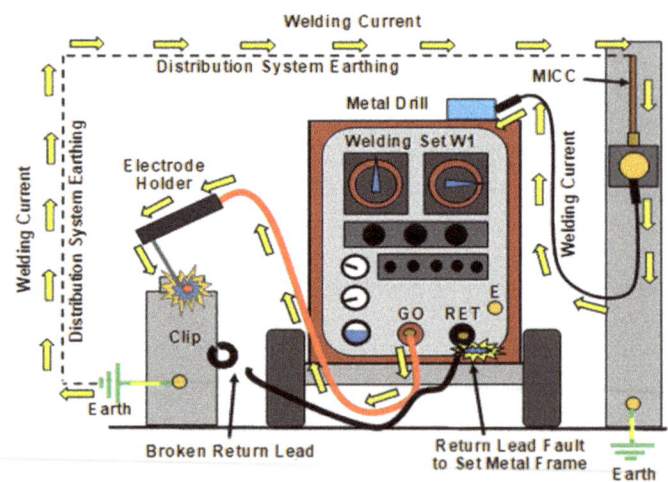

The distribution main wiring, including earths were in trunking and would not take the **80A** and so burnt the earth wiring out. In doing so, the melting wire destroyed the wiring around it.

The same can happen if there is a thin wire earthing the welding set, so a factory **'Electrical Standing Order'** came out stating that welding generators are not to be earthed to the plant. This actually caused a problem in later years as these generators started to include 110V generators tagged on the side where the welder could plug in a grinder. My argument of the day was; how does this smaller generator get its protection if it is not earthed to the plant? We changed the Standing Order!

1) Do you earth Welding Generators on your plant?

Large Portable Generators:

It must be ensured that the portable generator to be used on our chemical factory is TN-S or TT. The set must not be allowed to be TNC or TNC-S which I am sure the company that is loaning the equipment will ensure.

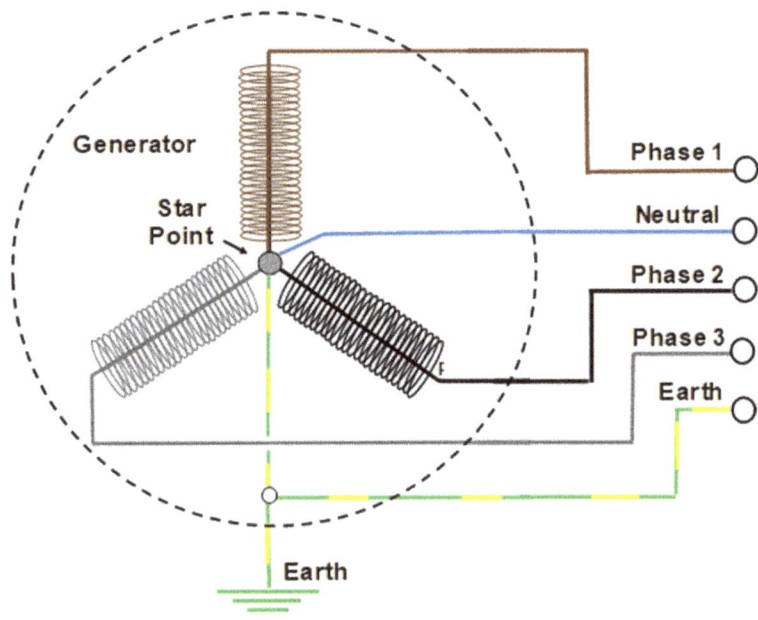

There must be no difference in potential between the generator Earth Output and True Earth otherwise we could get sparking. The neutral star point is earthed for safety by an insulated earthing conductor. Sometimes the generator need NOT be earthed but the manufacturer should stress this in the documentation.

The TN-S earthing system is one of the older systems, but safer for our Hazardous Area. This system does not rely on the earth block being supplied off the system neutral so the block should be at the same potential as the star point earth at the distribution transformer. The earthing is supplied from the electricity generator to the consumer via a separate core/**SWA of the feeder cable.**

On a **TN-S Earthing System** we have the **TN** part which is back at the Star Point Earth and the '**S**' because our earth wire at the consumer is Separate from the neutral. In this system the resistance/impedance would be much lower so a larger earth fault current may flow momentarily, but the protection would hopefully act quite quickly.

These large sets are usually on a fixed frequency meaning that the speed of the generator cannot be altered. In this country we have 50Hz (50 Cycles/Second) so we cannot speed the generator up to 60Hz if American equipment is to be used. Likewise an American generator cannot be slowed down to 50Hz if UK equipment is to be used. If the generator has earth protection relays or RCDs the documentation may insist that the generator is earthed to the plant to ensure the protection systems work in the event of a fault.

Many American welders have asked me when I was on plant if they could have 60Hz instead of 50HZ which of course we could not give them as we could not speed up the generator. We had to explain that in this country we are stuck with 50Hz unless they bring an American generator with them.

Chemical Storage Tank Earthing:

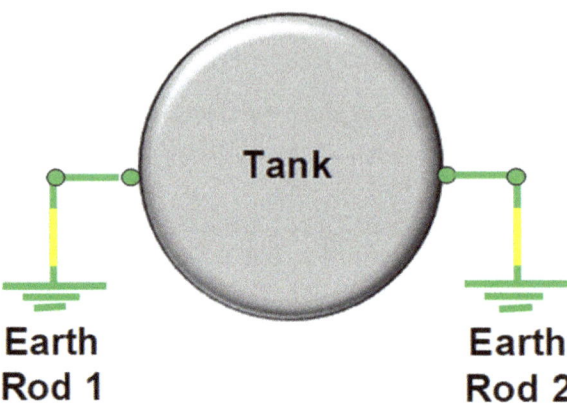

We have got to look at why we earth tanks.

1 – Is it part of the equipotential bonding system ensuring that all metalwork on a chemical factory is at the same potential?

2 – Well one reason is lightning protection. We do not have to have the lightning conductor connected to the tank, just erected near it, so long as the tank remains within the umbrella, otherwise we must erect one at the other side of the tank.

3 – The most important reason for earthing tanks, in my opinion, is to dissipate the static charge in the liquid if it is a hazardous material with a good insulating property such as toluene. Sometimes this type of liquid can have a long settling down time before the charge dissipates, maybe as long as 30 minutes depending upon the size of the storage tank.

There are usually two earths at least, so one can be disconnected for testing whilst the other is keeping the tank safe. The tank diameter might determine the number of electrodes. This of course is risk assessed.

The liquid can enter the tank in several ways, one of the most common being a fill pipe from the top to the bottom of the tank so that filling is into liquid and not splash fed. The last thing we want is a difference in potential between the fill pipe and the tank, because that is a recipe for a static spark so we bond the pipe to the bottom of the tank. The surface of the liquid has a static charge. The tank is earthed and the liquid is in contact with the tank wall so the charge will very slowly dissipate through the earthed wall and base. Floating instrumentation and level indication can cause differences in potential so manufacturer's advice is essential here.

The more electrically conductive the liquid, the faster it will lose its static charge. Tanks can be connected to an earthed network or have individual rods. If it is a network it would be pointless disconnecting the tank earthing connections and testing, as you would be testing the full network and not an individual rod so the reading would always be a very good one.

1) Have you got large Storage Tanks on your plant?

2) What Earthing Systems are in place?

3) What is the conductivity of the liquids that are in the tanks?

Earthing Chemical Storage Tanks:

On a chemical factory the tanks are earthed against static charges building up on the surface of the liquid content. Sometimes the tank earthing is provided via earth rods as below and sometimes via a tank farm earthing grid where the tank earths link into it. Individual tank earthing cannot be tested here of course.

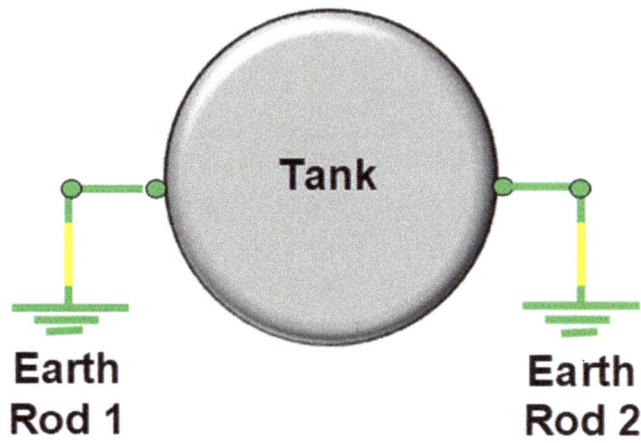

Left is the earthing on a chemical storage tank showing two Earth Rods so that one can be removed for testing and the tank is still protected. On very large storage tanks there may be three earth rods. These rods, as mentioned previously, are more for static charges in the tank contents than protection against fault current. All you are doing here is finding somewhere for the 'borrowed' electrons of static electricity to go without causing sparks. Remember the tank may also have 240V lighting.

Hydrocarbons by their very name are made up of a mixture of the elements hydrogen and carbon. Many hydrocarbons that we are interested in come from oil which, in itself, is a hydrocarbon. Many of these liquids are very bad conductors of electricity and hence ideal for both producing and holding electrostatic charge. In the bonding of pipeline section we discussed how a pipeline might become charged by the substance passing through it.

The liquid can enter the tank in several ways, one of the most common being a fill pipe from the top to the bottom of the tank so that filling is into liquid and not splash fed. The last thing we want is a difference in potential between the fill pipe and the tank because that is a recipe for a static spark so we bond the pipe to the bottom of the tank. The surface of the liquid in the diagram on the right has a static charge as indicated in red. The tank is earthed and the liquid is in contact with the tank wall so the charge will very slowly dissipate through the earthed wall and base. The more electrically conductive the liquid the faster it will lose its static charge.

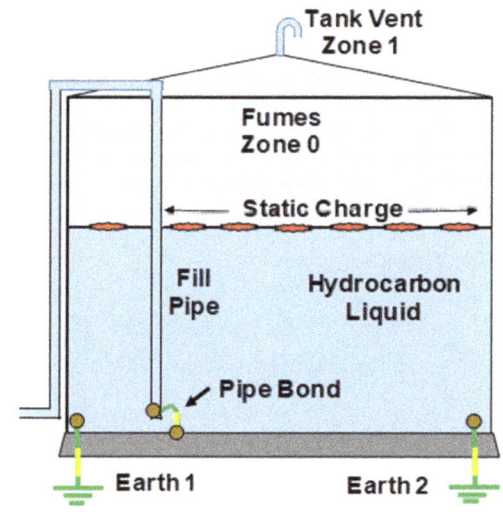

Tanks can also be fed from the bottom, which again will fill into liquid, but splash filling should be avoided as this will only serve to enhance static charge. If the tank has a stirring mechanism causing liquid turbulence this may also enhance the static charge.

This action of liquid losing its static charge is called 'Relaxation' time and in some liquids can be quite a few minutes. For instance a large 10,000 gallon tank may take around 30 minutes relaxation time, a 5000 gallon tank around 5 minutes and tanks under 5000 gallons around 1 minute. These values are risk assessed depending upon the conductivity of the liquid.

Earthing and Bonding Floating Roof Tanks:

Floating roof tanks, as far as storage tanks go, are not common. We had one on our Naphtha storage tank on our chemical factory. They are usually used where the chemical has a high evaporation rate. The diagram below shows the way that the tank may be earthed and bonded.

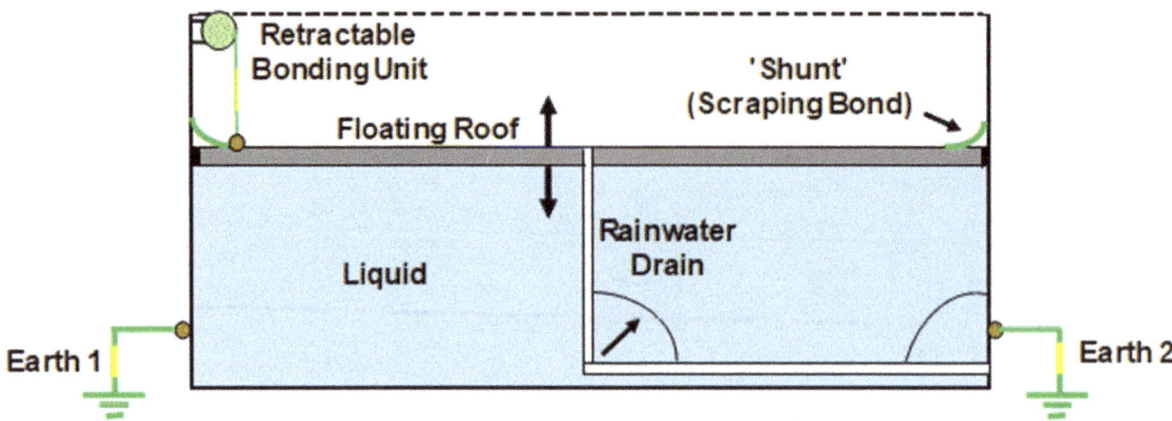

So the tank itself i.e. the walls, were earthed with two or more earth rods, as any other storage tank would be. The diagram above shows that the tank is earthed with two earth rods, so we earth this tank mainly for hydrocarbon liquids to dissipate their charge after a relaxation period.

The earthing also offers a degree of lightning protection for the tank. The last thing that would be needed is for a lightning strike near to the tank to charge up the outside to a different potential to the roof and the liquid!

The problem here is that the floating roof has seals around the roof itself otherwise the contents could escape upwards. This in actual fact makes the floating roof metalwork independent of the tank which, of course, opens the door to the roof becoming statically charged independent of the tank.

The way the roof was bonded in the past was to fix what are called 'Shunts' on to the roof that scraped the sides of the tank as the roof went up and down, but after a few years these become strained and do not scrape as close as they should due to constant use and wear. If the roof for any reason became independent then of course it is dangerous.

One of the modern ways of bonding the floating roof to the tank wall is to use a retractable bonding wire which is similar to a washing line. As the roof moves up and down the bonding wire unwinds and winds from the drum.

I have shown one in the diagram above but a large tank may have several, depending upon manufacturer's guidance. Apparently these retractable bonding leads can be obtained with Atex approval.

1) Have you got any Tanks with Floating Roofs?

2) If YES how is the Floating Roof Earthed?

Variable Frequency Drive Earthing:

Let me first set the scene and explain what a variable frequency drive is and then you might understand the earthing problems involved. All Variable Frequency Drives (VFDs) control the speed of an AC induction motor by varying the motor's supplied voltage and frequency. In the past motor speed could be varied by voltage alone by an 'Auto Transformer' type system, but drop the voltage too much and the motor torque will be affected leaving the motor with not enough torque to turn the load.

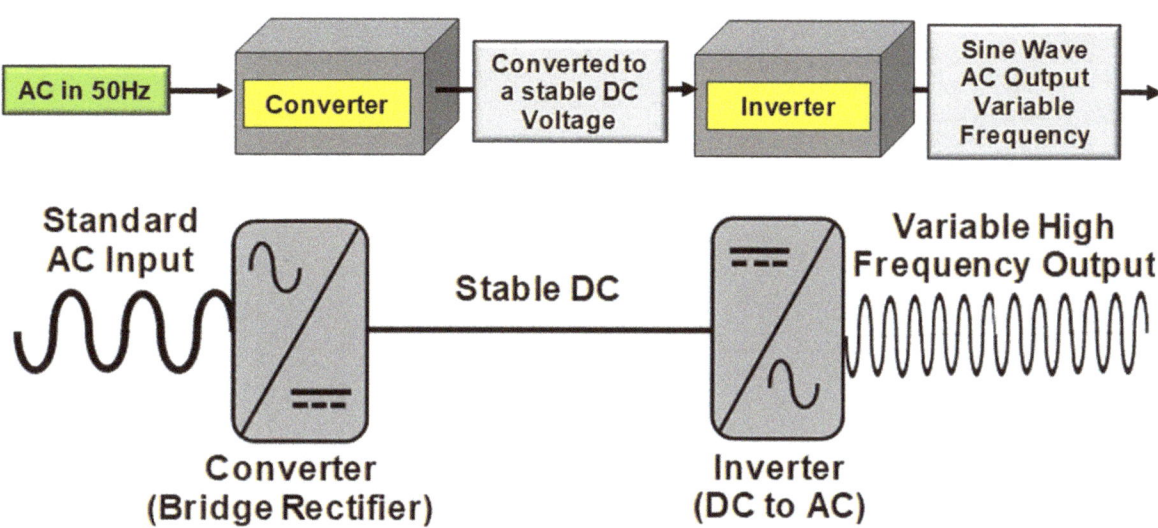

With the Variable Frequency Drive, basically what they are doing is feeding the AC line supply voltage into a 'Converter', as above. This basically is a huge bridge rectifier that changes the AC input to a stable DC voltage. The DC voltage then passes into an 'Inverter' which will then regulate both voltage and frequency and create a sine wave output to the motor.

The most common and lowest cost system seems to be the **Pulse Width Modulated System (PWM)** which has variable switching frequency from 1KHz (1000Hz) to **around 20KHz (20,000Hz)**. With PWM it is difficult to balance the phases of the output to the motor and it can end up with what is called **'Common Mode Voltage' (CMV)**.

The common mode voltage (CMV) induces currents into the rotor shaft causing massive bearing problems (mentioned under 'bearings') for the user as well as unwanted sparking possibly in Zoned areas. When these stray currents pit the bearings it is known as **'Electric Discharge Machining' (EDM)**. This EDM sometimes also attacks the bearing lubricants by creating high discharge temperatures thus carbonising the oils. The attractiveness here is that full torque can be achieved from the parked position to maximum speed.

Harmonics are a problem with these systems but they can be filtered out. A huge benefit for large companies which may have a lot of induction motors and transformers on site is that Variable Frequency Drives run with a Power Factor of around 0.8 so can be used for factory power factor correction.

1) Do you have any Variable Frequency Drives on your plant?

2) Have you looked for problems with stray earth currents?

Stator to Rotor Coupling Current:

I am going to talk about currents caused by variable frequency control here because it is the bearings of the machine that can be dramatically affected. I have started with a pedestal motor where there are white metal bearings.

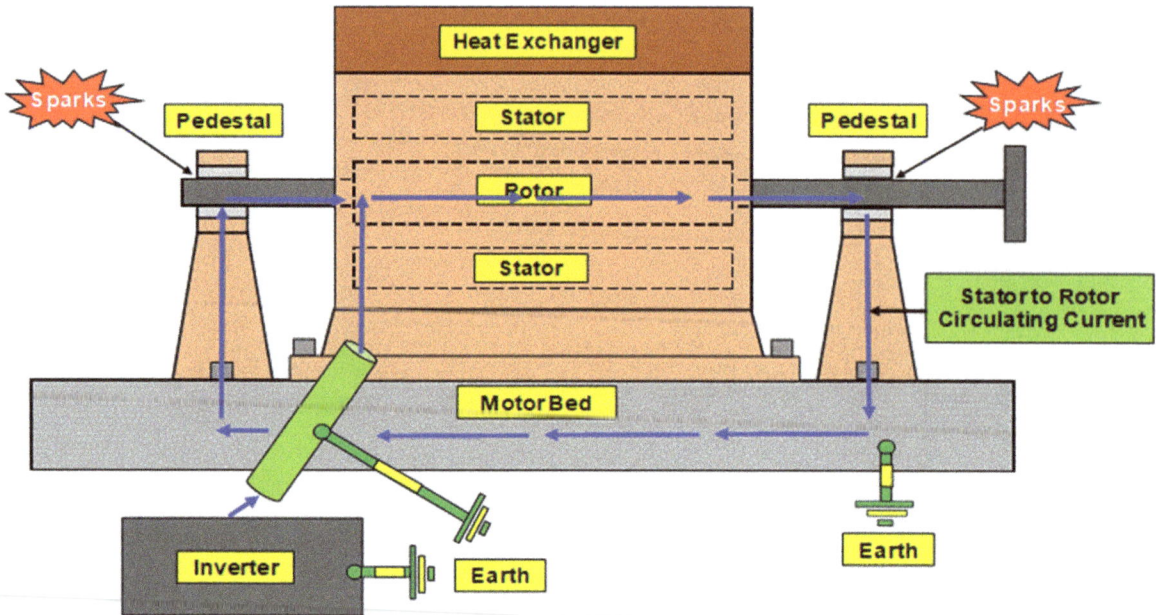

Stator to Rotor Coupling Current: How does the circulation current do damage? Well, let us take a larger induction machine where the rotor rests on pedestals with white metal bearings. The pedestals and motor frame are resting on and connected to a metal earthed bed, at the drive end and non-drive end. Capacitive circulation currents pass from the stator jumping the air gap into the rotor, along the shaft and down the pedestals and through the bed causing sparks to track across the white metal bearings in the pedestals. This could also happen to a motor with conventional ball or roller bearings which could be in your Hazardous Area. These motors could be used in a hydrogen compressor house which, of course, would be a gas group IIC area.

The Circulating Current is marked with blue arrows on the above diagram. How can we stop or slow down this type of current from causing sparking and pitting of the white metal bearings in the pedestal or the balls/rollers in conventional bearings? Stopping the circulating current can be done in several ways, some of which we can dismiss straight away because of cost and some we can dismiss because of safety in the hazardous area.

A – Insulate one pedestal away from earth. (Quite common solution for pedestal units.)

B – Ceramic coat the bearings (Not the cheapest way.)

C – Use bearings with ceramic balls or rollers. (Not pedestal bearings.)

D – External or internal Shaft Grounding Brushes. (Difficult in hazardous areas because of sparking.)

E – Insulate the bearing if conventional bearings. (Difficult, would require Manufacturer involvement.)

F – Sealed bearings that have been packed with electro-conducting grease.

G – Inversine Advanced Universal Sine-Wave Filter. (Probably the most costly.)

H – Converting to Magnetic Variable Speed Drive Couplings or Sheaves (New Technology.)
    (All require Manufacturer involvement/approval.)

Rotor to Shaft Current:

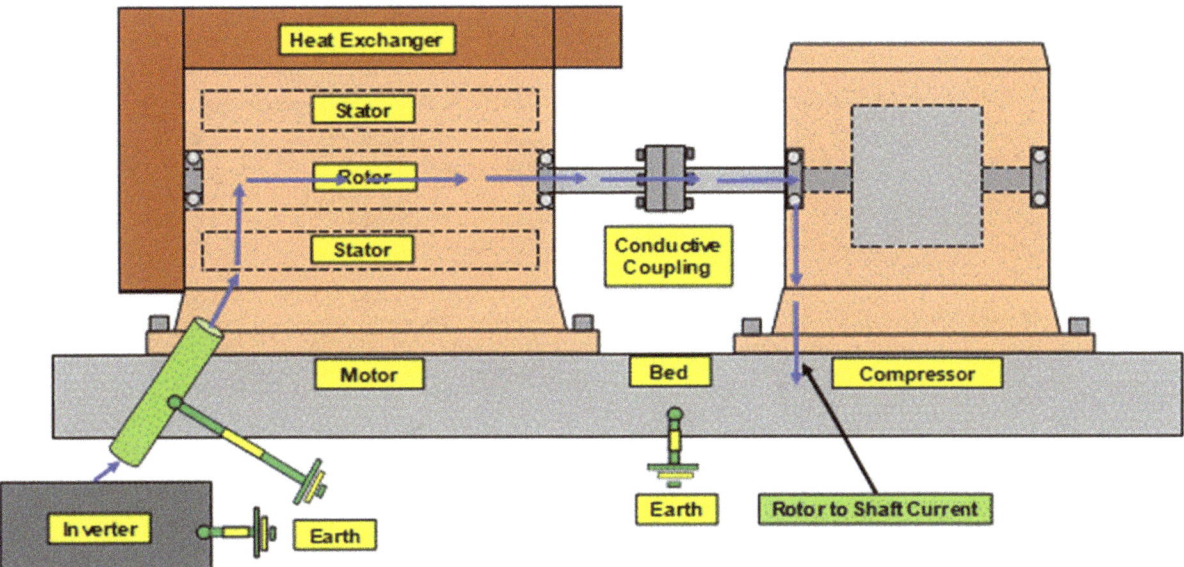

Rotor to Shaft Current: This particular current actually jumps the rotor–stator air gap and travels along the shaft, through the coupling into the load and through the load bearing(s) to earth.

So in this case the circulating current can actually affect the bearings in the driven unit. These shaft currents as mentioned not only put the bearings at risk in the load as well as the motor, but also cause unwanted sparking in our Hazardous Areas.

Stator to Shaft Current:

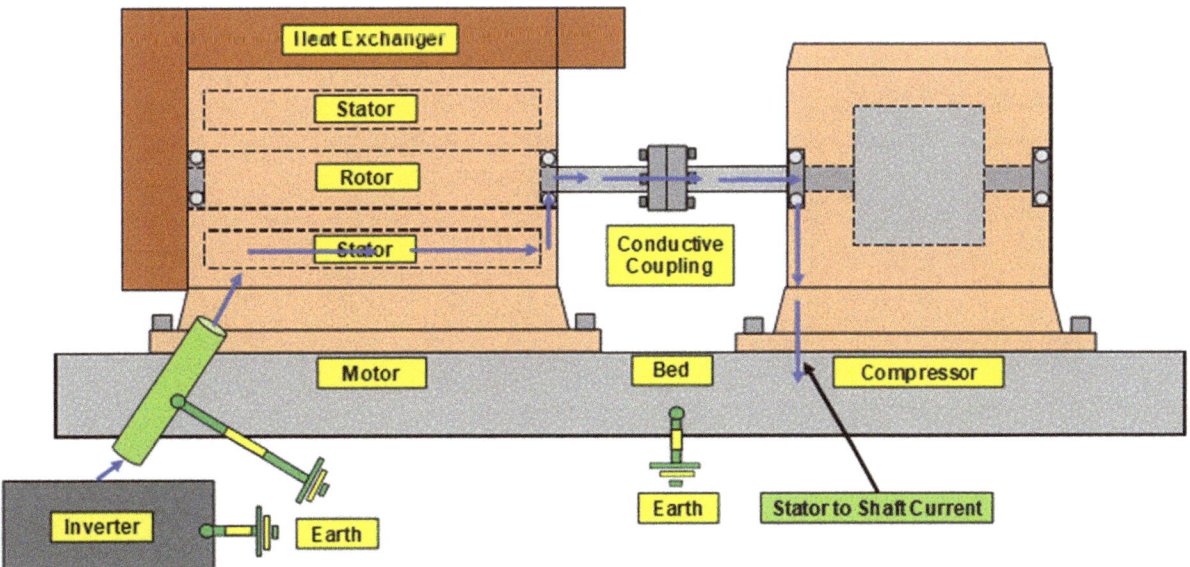

Stator to Shaft Current: Similar to the rotor shaft current except this time the current passes from the motor stator through the bearings on the motor drive and the driven unit on its way to earth. So here motor and driven unit bearings are at risk.

Stator to Ground Current:

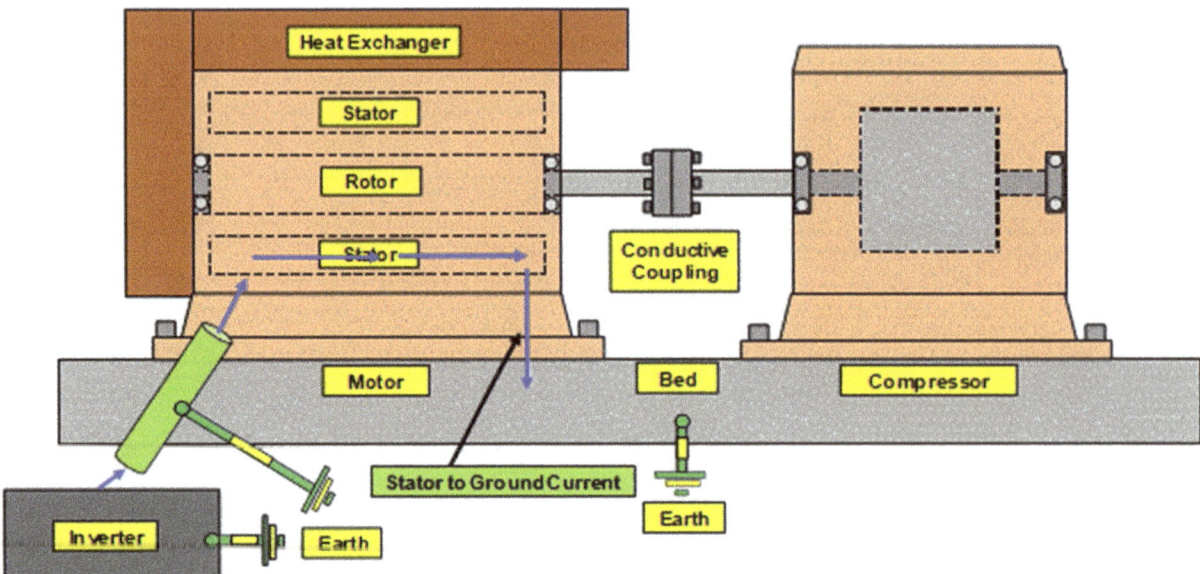

Stator Winding to Ground Current: This would be the ideal situation where any stray currents in the stator winding find their way to ground without going through any shafts or bearings, hence no sparking.

There are two phenomena that these electric currents may cause: 'Fluting' or 'Frosting' and I have given an idea below how they may manifest themselves.

Fluting:

When bearings start giving problems and have eventually to be changed and the faulty bearing is inspected sometimes there is a phenomenon called **'Fluting'** when the bearing race is inspected. These are usually little grooves parallel to the bearing axis that are caused by electric currents passing through the bearing. These grooves could be in just a small part of the bearing race or all the way round. The diagram on the right shows a section of bearing race and where the fluting may occur, mainly where the point of a ball or roller touches the race.

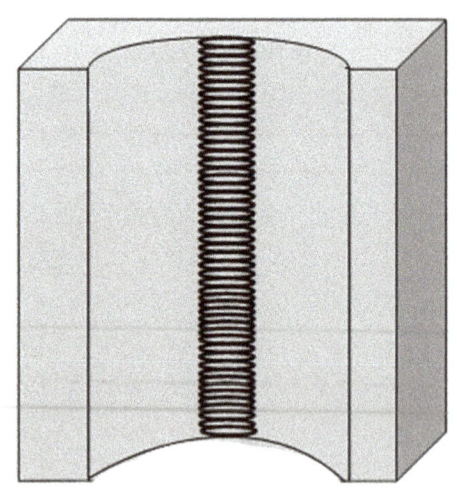

Frosting (Pitting):

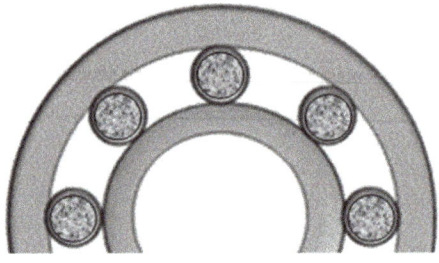

This is another indication of electric currents running through the bearings. When you inspect the bearing the balls and races have a dull appearance which is the **'Frosting'**, instead of shiny as you would expect. If you look closer at this frosting you will see thousands of little 'pit' marks making up the appearance. The diagram left shows the frosting texture on the balls.

Electrical Earthing Drawing Symbols:

There are many different types of Electrical Earthing. You may recognise some of the drawing symbols below and some may be quite new to you. Earth symbols are found in **IEC 60417**.

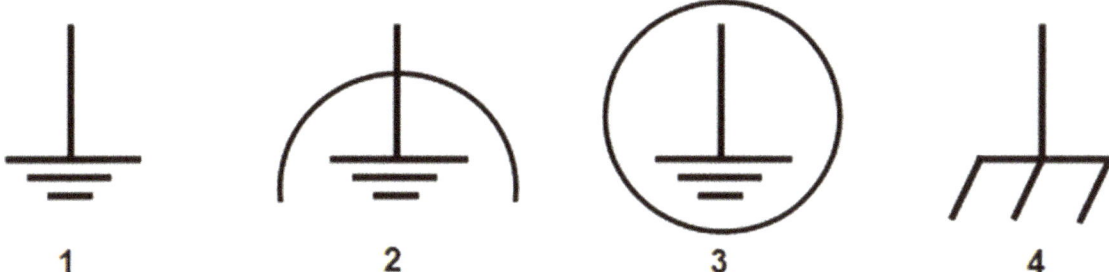

No.1 – This is probably the symbol that is most recognised. This is the symbol for a Standard Electrical earth. Three bars forming an arrow head.

No.2 – This symbol may not be quite as common and it is the symbol for a 'Noiseless' Clean Earthing terminal.

No.3 – This symbol identifies a terminal/point on some system that should be connected to a Standard Earth Point.

No.4 – This is a chassis or frame grounding point. Similar to No.3, but on a frame.

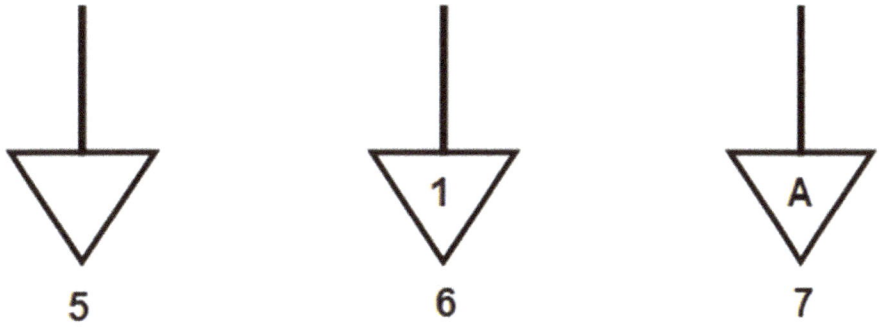

No.5, No.6 and No.7 – These are the symbols for 'Analogue' & 'Digital' Earthing if you ever come across them on a drawing. They may be seen on drawings where there are 'Integrated' Circuits where there must be no electrical noise.

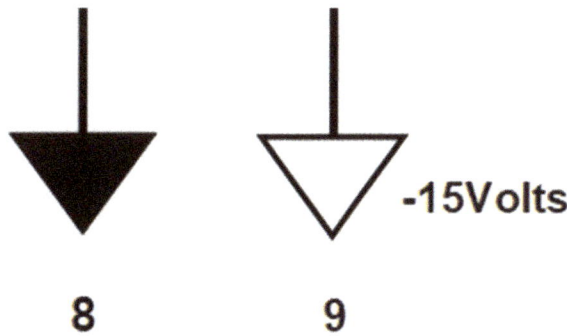

No.8 is called a 'Virtual' Earth off an op-amp. Anyone who studies guitar electronics may come across this one. No.9 is called a common earth with modifier. This earth will be common to all other earths on the drawing with the same symbol.

Double Insulation:

You might be wondering why I have included a system which does not require an earth in a book about earthing. Sometimes topics like this one make you realise what an earth actually is and why some equipment does not require one. This equipment with regards to PAT testing is of course **Class II**.

The 13A plugs in this country are 3 pin so in this case only the live and neutral pin is used. In some appliances the plug may well be sealed onto the cable so that you cannot remove it, you can only change the fuse.

Also in some double insulated appliances the plug earth pin may be made out of plastic so that the plug fits the socket. In normal plastic 13A socket outlets it is the shape of the actual pins that allow the shutters to open but in some cases the earth pin opens the safety shutters. This can be open to children poking things into the earth of the socket and opening the shutters. Another thing that I will mention here is that the live and neutral pins of a 13A plug are insulated near to the plug as it was possible for children's thin fingers to have contact with these pins as they plugged in.

Do not for one minute think that all double insulated items are made out of plastic. I can obtain double insulated lighting Class II that is made of similar material to Class I. It is the way the insulation of live conductors is arranged that makes them double insulated. One thing that I might mention here is that double insulation seems ultra-safe, but let me tell you that it becomes very dangerous if it gets wet. People mistakenly believe that double insulated appliances can be used in a bathroom for instance.

People have been electrocuted in their bath by using their mobile phone with their double insulated charger plugged in. If water penetrates the outer cover of the unit then the whole outer case could come live. Usually there are no fuses on these charging units to blow, only the fuse in the 13A plug. People still use double insulated lawnmowers, electric drills & hedge trimmers etc. when it is raining which is extremely dangerous. If the area is wet and water could get onto the electrical equipment or the weather indicates that it could rain, do not use double insulated tools.

It is not advisable to make up extension leads etc. just for double insulated tools **with just a live and neutral in. YOU MAY GET THEM MIXED UP AND END UP WITH METAL EQUIPMENT NOT EARTHED!** If you make extension leads up for double insulated equipment, make them 3 core not 2 core and then they will not get mixed up with the normal ones.

1) Have you got Double Insulated Equipment on your plant?

2) What procedure do you carry out when PAT Testing?

In the past, tools such as 240V electric drills would have a metal case which for safety would require earthing by means of an earth wire from the plug. These days 240V tools are usually 'Double insulated' and only require a live and neutral from the plug with no earth. The idea is that if the equipment had only one layer of insulation and it had a flaw, voltage may escape, but this is unlikely with two layers.

The mark for double insulation is left i.e. a box within a box and will be shown on all tools and equipment that fall into this category such as electric drills, hairdryers, vacuum cleaners etc.

This equipment when PAT testing would be Class II. During PAT testing this type of equipment can get away with a visual inspection unlike Class I equipment which would have to undergo the full testing procedure.

So what does double insulation actually mean? Well first of all the case is made out of re-enforced plastic so any faults to the case will not be a problem. Two independent insulations would have to fail to cause a problem.

There is a lot of talk about meter tails being double insulated as they have one insulation on top of another. This, **in my opinion**, is not so as they are no more double insulated as twin and earth and they are not exactly equipment in their own right.

Internal electrical components are in their own insulated casings making it even more difficult to cause a problem, i.e. the switch, motor etc. One problem with this protection is that if, say, there was a primary 'Live' core earth fault no one would know and then if there was to be a secondary fault on the 'Neutral' core and they both shorted there could be a fire. Now if PAT testing does require a 500V insulation test, well all that happens here is that the appliance is plugged into the PAT tester and the probe is put onto something metal on the appliance such as screws, the chuck on the drill etc. with the appliance switched to on.

Now of course this type of equipment is **NOT** usually Ingress Protected and would become extremely dangerous if it got wet. If water penetrated the outside case and leaked onto live parts then, although water is not a very good conductor. it will conduct and the case could become live. Saying this, people still operate electric hedge trimmers and lawnmowers in the rain! Another very dangerous trick that people do is to plug the double insulated charger into their mobile phone and use it in the bath. The phone might be ok but the charger can cause serious electrocution, leading to injury or death.

UNDER NO CIRCUMSTANCES MUST YOU MODIFY THIS EQUIPMENT TO INCLUDE AN EARTH AND YOU MUST NOT DRILL THE CASE IN ANY WAY!

There is another type of insulation similar to double insulation and that is 'Reinforced' Insulation. This improved insulation provides the same protection as double insulation against electric shock. Protection is still provided should the basic insulation fail. Instead of two separate insulations i.e. motor in a case within a plastic outer case (double insulation, box within a box) this equipment has a very strong reinforced case.

Water Jet Cleaning and Cutting:

Water Jet Cleaners and cutters range from small portable units which are little larger than a vacuum cleaner to large industrial units that are mounted on the back of a truck. You may have a smaller size unit yourself at home to jet wash your drive and car. We are going to look at the larger industrial jet cleaners and some of the problems that may crop up. There are several safety risks which I will mention, but our real concern here is earthing and do we do it?

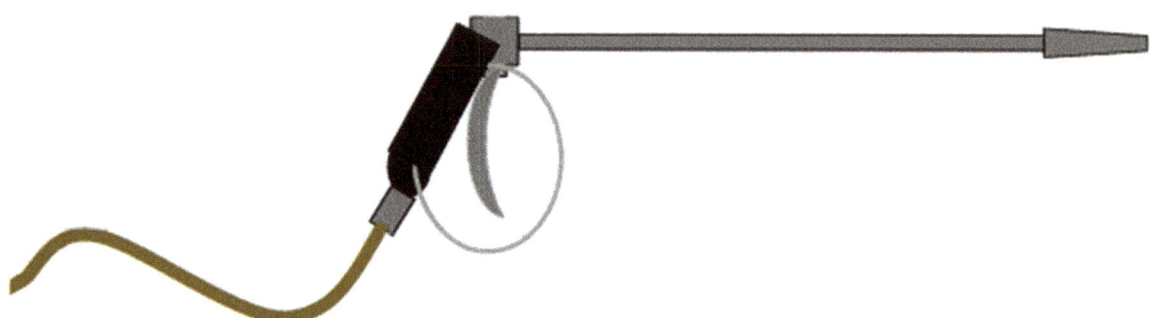

The company that comes onto our chemical plant to carry out the jetting, or in some cases cutting, turn up with a large truck type as mentioned above with the unit mounted on the back. The jetting or speciality cutting nozzle may be able to produce water at around 1500Bar (21,000lb/square inch). This pressure will cut through human tissue very easily so the company will ensure correct PPE and competence. Pipe cutting can be done with the water jet at around 3000Bar (around 44,000lb/square inch.) so the company should again ensure PPE and competence.

I read an article on water jetting when I was with BP Electrical Department that caused us some concern. We hired a jetting unit and by the above figures you can guess that this operation was for very high pressure water jets to be used on tank cleaning. Apparently, jetting can, in certain circumstances according to this article, **CAUSE SPARKING.** The danger is that there can be a **20KV** static charged mist that is generated by the water pressure and as the water passes through the mist a spark is produced.

Now, apparently, when they jet clean tankers it is done in an atmosphere of inert gas and according to the article, the sparks that are produced are incendiary and quite capable of igniting a gas.

Before any operation of this magnitude is undertaken, I suggest that a risk assessment should be completed by the Electrical Engineer as to how far you go with earthing and whether it will prevent the sparking.

The risk assessment might ask: Do we just earth the unit on the truck for instance, or would it do any good to have an earth wire running along the pipe right up to the jetting unit head?

1) Does your plant have Jetting Contractors and a Policy for earthing?

2) Did you know that there may be a problem with Static?

3) At this moment in time do you earth up to the Jetting Nozzle?

Steaming Out Bay:

So why have a steaming out bay in the middle of our chemical factory and what problems arise with the steam? In the procedure below are we Earthing or Bonding? Maybe a bit of both in this case.

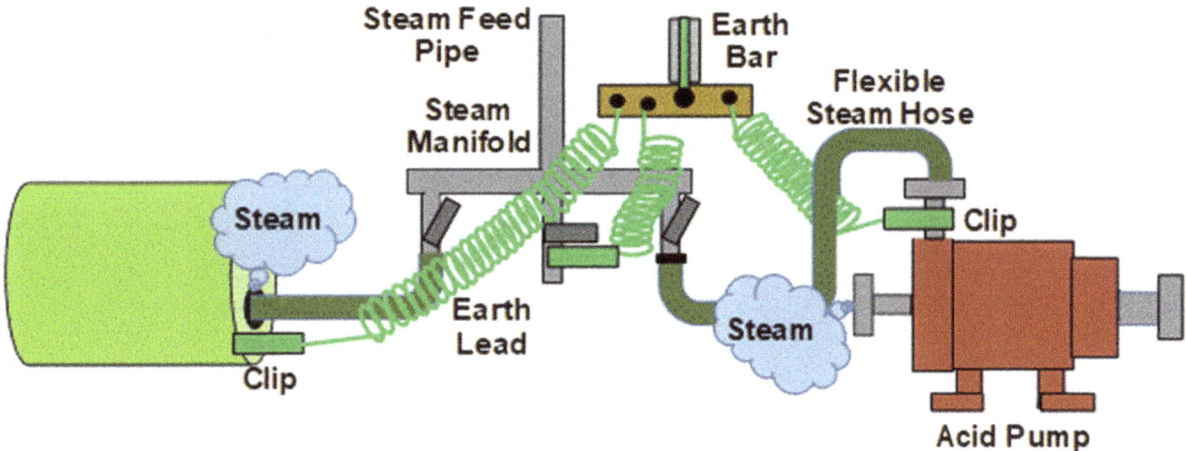

Well firstly let us look at the diagram above and talk about the different parts. You will see that we have an acid pump and a drum which may have had some hazardous chemical in it. You will see that there is a steam manifold and from the manifold there are flexible steam hoses that go to each of the items of equipment being steamed out.

Why do we have this system? Well if we take the mechanical workshop, the lighting, sockets and electrical equipment is a little bit more robust than domestic, but still non-certified. We intend to remove the above acid pump from site and transport it to our workshop. The pump is still full of acid and when opened in the workshop could turn it into a hazardous area with fumes etc.

Firstly we would take it to the steaming out bay and put a flexible steam hose into the top and the steam will come out the bottom. We can angle it if required. The steam will wash out the inside of the pump and ensure that there is no acid within when it arrives at the workshop. An untearable steaming out permit can be attached to let the workshop know that it is chemical free.

Now steam is known to produce static electricity and could charge the various items of equipment being steamed out to around 10,000V as independent items. How do we counteract this? Well we have an earth bar with multiple cables and we connect one of the earth cables to the pump and give it an equipotential bonding.

This is why very vicious steam leaks should be repaired as quickly as possible as they are a 10,000V spark risk. Coiled Earth Leads are made these days, which ensure that everything does not get in such an untidy state as it did using catenary wire with clips fastened on and the leads now coil themselves shorter and look much tidier.

1) Do you have a steam cleaning facility?

2) If the answer above is YES is it suitable?

3) Are items always earthed properly in the bay?

Socket Polarity and Earth Tester:

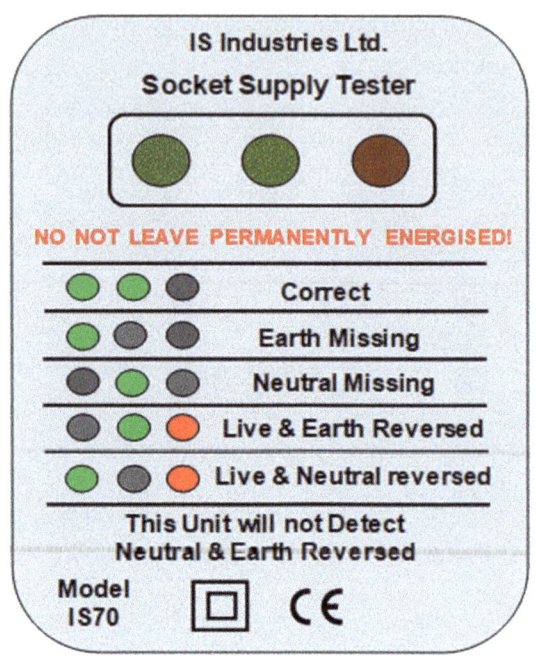

Socket polarity testers are used for a number of reasons as in the diagram, left, but two of the important tests that are carried out as far as this earthing book is concerned are:

1) Is there an earth present at the socket? and

2) Have the live and earth been crossed in the socket connections?

Depending upon the Electrical Engineer, sometimes this is tested in our hazardous area before, say, **EVERY** shutdown and sometimes on routine maintenance.

The lights at the top illuminate accordingly.

To test the sockets in our Hazardous Area an adaptor is made to fit the Exe Increased Safety sockets on site such as the one below, with a certified Exe plug going to a 13A metalclad socket.

Of course due to the 13A socket this equipment would not be certified Exe anymore so a label must be fixed on the equipment explaining this.

Also a **GAS FREE CERTIFICATE** must be obtained on a running plant.

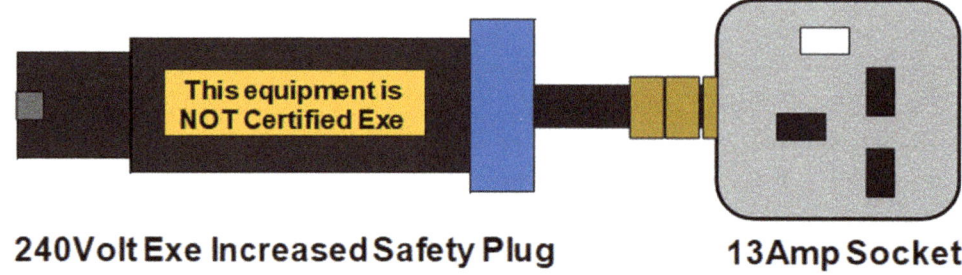

Now the test unit at the top can be plugged into the 13A 240V socket above and the tests can be carried out. The tester will not of course inform if earth and neutral are crossed because as far as it is concerned they are the same thing.

Just several safety reminders:

1 - Under **NO CIRCUMSTANCES** must this adaptor be left on plant. Remember it is not certified.

2 - Under **NO CIRCUMSTANCES** must this tester at the top be left permanently plugged in live. The unit at the top is not designed for permanent use.

3 - A **GAS FREE CERTIFICATE** must be obtained before this device can be used in a **Zoned** Area. (As per company policy.)

1) Does your plant check their socket outlets?

2) Do you have adaptors similar to the one above?

Bonding:

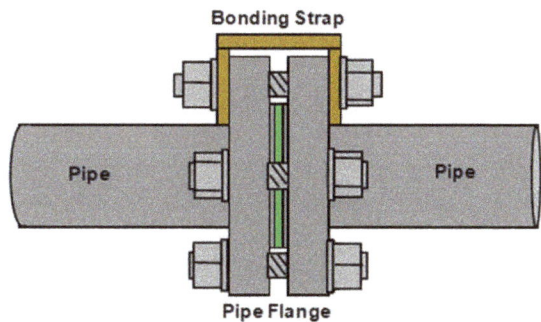

It has to be understood that Earthing and Bonding are two different things. Earthing means something is wired so that it is in direct contact with earth via earth rods, plates, bars connected to rods etc. Bonding is an indirect system that leads eventually to earth. It is a procedure that ensures that most metalwork on the factory is the same potential **(equipotential)** just as it would be in domestic premises with water and gas pipes etc. so the risk of different potential sparking or shock is low.

There are two types of bonding i.e. **Main Bonding** (similar to direct earthing) and **Supplementary Bonding**. Main bonding is where, for instance, we take a wire from an earth bar to pipework or vessels. Supplementary bonding is where we take a wire, say, from one vessel to another.

Relying on nuts and bolts for continuity is not a very good way to achieve a good bond. **Rust for instance is an insulator** and I am sure that you cannot guarantee rust free conditions on a chemical factory even with newish installations. Whatever you use to stop or slow down the rust will sometimes result in more insulation. The next pages give examples of bonding systems and instances. The following examples of items require a bond wire/strip/braid. I am sure you can think of more:

1 – Exd Flameproof pieces of equipment – Bonding screws on items.

2 – Exe Increased Safety pieces of equipment – Using earth tags.

3 – Exn Reduced Risk pieces of equipment – Using earth tags.

4 – Cable Tray – Sometimes **across each joint** and one point to the structure.

5 – Ladder Rack – Sometimes **across each joint** and one point to the structure.

6 – Vessels and Columns on the plant.

7 – Storage tanks if they are on plant (tank farms slightly different).

8 – Pipe flanges if say in zoned areas. (Risk assessment & depends upon liquid.)

9 – Portable pumps.

10 – Tankers – Road & Rail. Sea Tankers may be different.

1) Do you bond Cable Tray Joints?

2) Do you bond Pipe Flanges?

3) If the answer above is YES then is this in Zone 1 Areas only?

Earthing and Bonding?

We need to understand what **Bonding** is as opposed to **Earthing**. Many people, including Electrical Technicians, think Earthing and Bonding are the same so what is the difference? Some Technicians even call it Earth Bonding which adds to the confusion.

Let us look at a **TN-S** electrical system. The Earthing Conductor is 'Separate' hence the **'S'** and this then would be the **Earth** and would be the path back to the Star Point of the Distribution Transformer should there be an Earth Fault at the Consumer, possibly through paths such as the steel wire armour of the supply cable.

Let us take a metal socket outlet on the end of an extension lead. The metalwork of the socket must be **Earthed** to the **'S'** of the **TN-S Earthing System** to stop you getting an electric shock in the event of a fault to the case. What must be ensured now is that all metalwork not directly part of this **'path'** is connected to the Earthing point to ensure that there can be no differences in potential should a fault occur. So the metal case of the socket outlet would be called an **'Exposed' Conductive Part.**

Let me use two examples of Cable Tray and Ladder Rack. These are what are called **'Cable Management'** and it is the way cables are run through the plant. Cables and electrical equipment can be fixed to this metalwork of the tray, but it is not a direct part of the Earthing System and as such is not meant to carry fault current, so at some point would have a **Bond** of its own. Therefore the cable tray would be called an **'Extraneous' conductive part.** The above-mentioned cable tray would be an ideal **Equipotential Bonding** situation. If we fixed a metal socket outlet or switch to the cable tray we would run a **bonding wire** from the metal socket or switch case to the metal work of the cable tray just as you would, in a house, take a bonding wire to all pipework.

Now let us take some more of the metalwork in our Hazardous Area which we would call **Extraneous Conductive Parts** and may require an **Equipotential Bond**: columns, vessels, pipework, metal girders of the structure. Some of these are discussed in their own right further on. We could easily get mixed up here with **Static Earthing** on, say, large storage tanks which may be installed for a different reason.

So the **Exposed Conductive Part** could become live if it was not earthed in the event of a fault to earth and the **Extraneous Conductive Part** could be other metalwork you could touch to receive an electric shock.

If we **Bonded** the two together **there would be no difference in potential.** Remember that in our Hazardous Area we are not just concerned with human protection, but with sparking as well. Two different items i.e. metal socket outlet and cable tray could spark momentarily in the event of a fault current to earth if they did not have an **Equipotential Bond.**

The diagrams on the next page may explain this problem a little better and you will see in diagrammatic form why we install this bond.

1) Do you think your plant bonds equipment correctly?

2) Are all Exe Junction/Marshalling Boxes earthed with 4mm cable?

Equipotential Bonding:

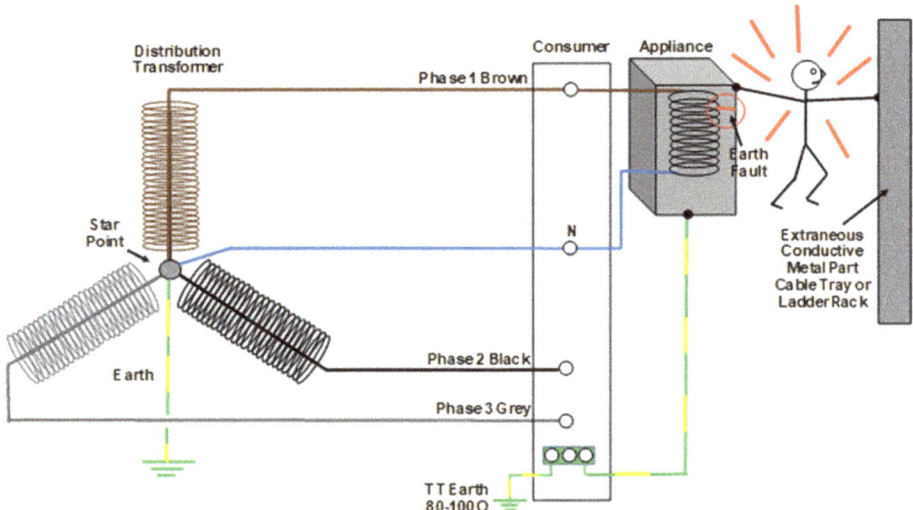

If we now look at what we discussed on the previous page in the above diagram, we need to bond equipment together on site because if we do not then a situation may occur where there could be a fatal shock. So let us say that our appliance is **NOT** bonded to the structure or the cable tray that it is close to, and an earth fault occurred to the metal case.

Now let us say that our rod on the TT system rod is around **80Ω–100Ω Impedance**. We could end up with a situation where the case of the equipment is at a different potential, short term, to the cable tray which is **Extraneous Metalwork**. So if our stick person was in contact with the cable tray and the outside metal of the equipment at the same time as the fault occurred, they could get an electric shock, which could be fatal. It is therefore imperative that we get the Earth Rod Impedance as low as possible.

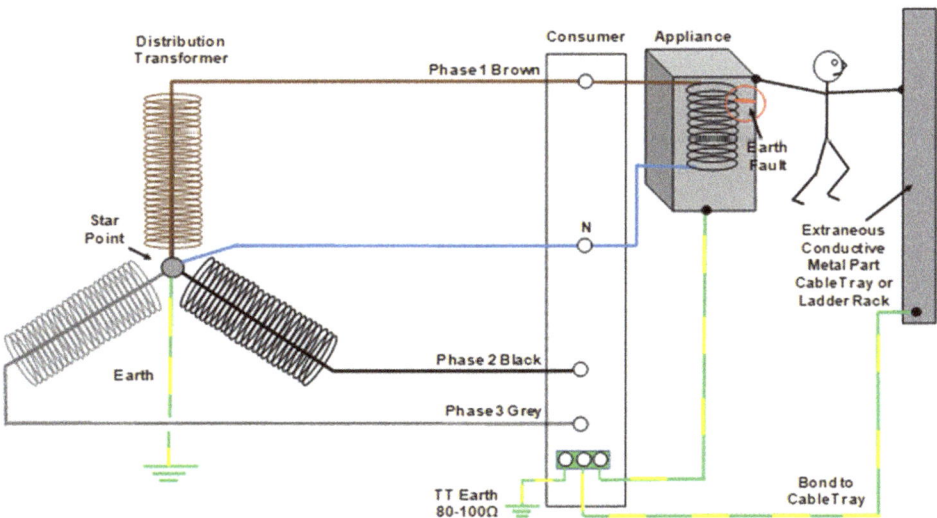

In the diagram above you will see that I have now taken a bond from the appliance to the cable tray via an earth block, so now we have made the cable tray the same potential as the case of the appliance. Our stick person does not now receive an electric shock unless they are a better earth than the TT earthing system. This would apply if you fixed anything metal to the cable tray, only then you would take a bonding wire directly from the appliance to the metal cable tray. This is described as **Equipotential Bonding.**

Exe Increased Safety Junction Boxes:

There are two junction boxes shown below. The top box has what they call an **earth plate** (called a 'crucifix' plate because of the 'cross' shape) which is really a bonding plate, and the bottom junction box has no plate. The diagrams are junction boxes, but below would apply to all Exe equipment.

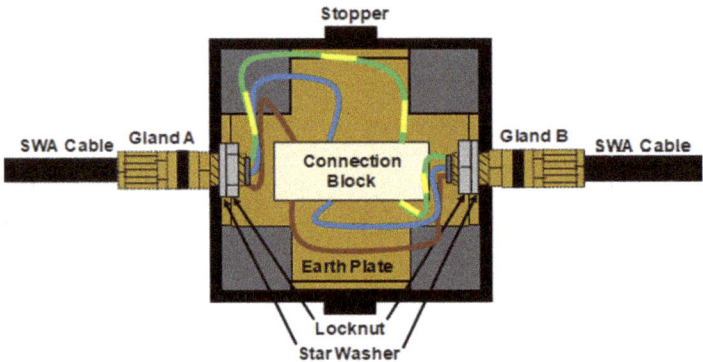

The above Atex junction box is usually a common sight in Hazardous Areas. This one has what is called an Internal Earth Plate or as above, Crucifix Plate, the aim of which is to bond together all of the glands that are inserted. In the above case, that is gland 'A' to gland 'B'. Although nothing is actually laid down, thought must be given, if mounted on a cable tray, to one of the glands having an earth tag and **4mm wire** to the metal tray. Of course a locknut & star washer is required to join the gland to the earth plate.

In the diagram above a three-core cable has been used with an earth core. This must always be considered instead of using the steel wire armour as the earth, which, may I add, is quite legal at the moment.

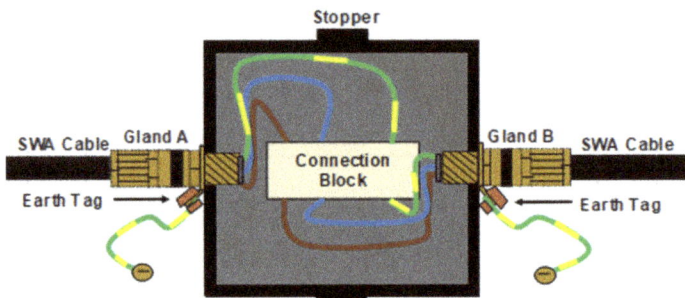

If the junction box has no earth plate the glands that are inserted into the box have to still be bonded together. This is done by putting an earth tag on **EVERY** gland and then joining all of the earth tags together with a **4mm bonding wire** and, if applicable, taking one wire to the cable tray or taking each individual **4mm wire** to the cable tray. If the box is mounted on a wall then the gland earth tags must be joined to each other and earthed at the earliest opportunity along the route.

Earth tags should not be fitted inside the junction box. There are other accessories.

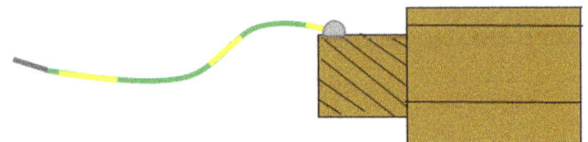

If it is deemed that an earth is required inside of the box from the gland then adaptors can be obtained 20mm–20mm with an earth wire soldered on.

Exd Flameproof Junction Box:

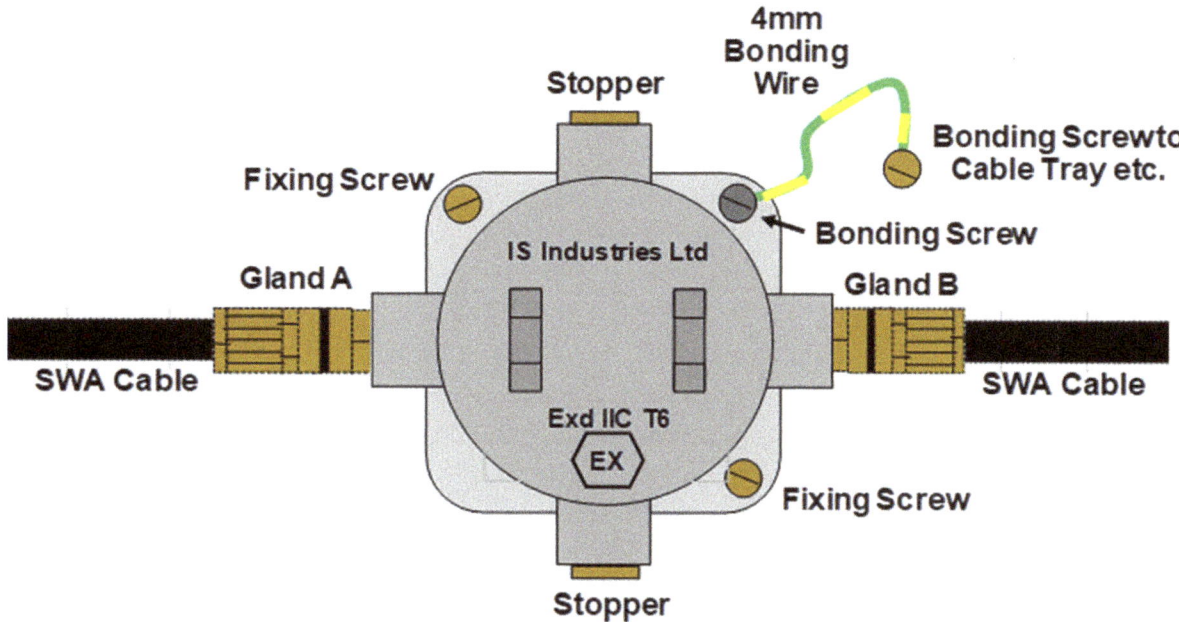

Flameproof equipment (Exd) is used less and less on chemical complexes as the increased safety equipment (Exe) is made of plastic and is therefore lighter, and maintenance is much easier. However there are many items of flameproof equipment still in use, and many Engineers will replace like for like

Above is a diagram of an Exd Flameproof junction box. This could well be a light switch, junction box or any other flameproof equipment. Because the equipment is now completely made out of metal this will automatically bond gland 'A' and gland 'B' together **without the use of earth tags.**

If it is company policy that earth tags be used on flameproof equipment then by all means put them on and also use an earth tag if by some chance the equipment has not got a bonding screw. Although this is not to do with earthing, some company policies state that fibre washers are used on **ALL** glands. Fibre washers should not be required on flameproof equipment, especially if silicon grease is used on the threads, but there is no legislation stating that they cannot be fitted.

On this equipment there are usually two fixing screws and somewhere a bonding screw. If fastened onto, say, a cable tray then a **4mm Bond** must be connected from the equipment bonding screw to the cable tray. **Do not under any circumstances use any other screw to connect the bonding wire to such as a flange or lid screw as this could compromise the certification of the enclosure.**

There may be an exception to the above if the equipment is fastened onto, say, a brick wall. You may have to run an earth strip along the wall. **Seek manufacturer's advice on instances such as this.**

1) Are all items of Exd flameproof equipment bonded on your plant?

2) Is the bonding wire 4mm² minimum?

3) Do you use earth tags with Exd equipment?

4) Have you got any Exd equipment mounted on a wall?

Bonding Flexible Hoses:

Static charge build up can be a huge problem in Hazardous Areas such as chemical plants, petrol stations and platforms. Anything that will build up and hold a static charge is a huge risk that has to be addressed.

Un-bonded flexible hoses pose a huge risk when loading or unloading hazardous flammable liquids or powders. The movement of the liquid inside of the pipe in contact with the pipe wall causes the charge. The following text discusses how we might reduce or eliminate the static charge in flexible hoses.

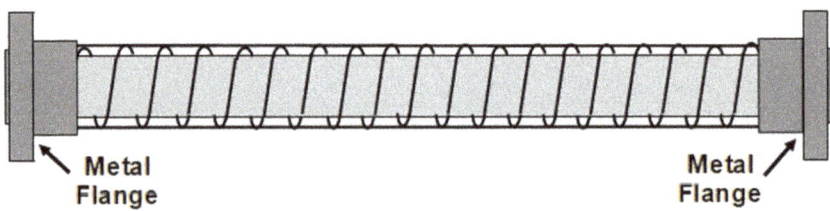

Flexible hoses are used on a chemical plant in many instances. If the plant has a jetty and facilities to load and unload oil and chemical tankers then these hoses will be in abundance at the jetty head. Road and rail tankers will also require these hoses, of smaller design, to load and unload. As you can see in the above diagram the flexible hose has a static bonding wire joining each of the flanges together. The bare wire coil is coiled around the hose and is in contact with the pipe material for the full length of the hose. Sometimes the wire is built into the hose and sometimes on the outside.

The hose could also be made of a type of rubber that is antistatic. As each hose is connected together the bonding is transferred along the hose sections and will eventually connect to a static pipeline which of course will be land earth. Bonding wires can sometimes be seen on the outside of the pipe and of course should be bare wire. Hoses must be individually numbered and must undergo regular pressure testing and continuity checks from flange to flange as advised by manufacturers and company policy. Please remember that in the case of a **sea tanker**, if we take that analogy, it is an electrode in an electrolyte (sea water) and when it sails in to the end of a jetty it may be at a different potential to land earth. Connecting bonded or metal hoses directly to the ship may cause sparks, the very thing our wire coil in the hose is trying to avoid.

In the old days we would connect a large lead from the ship to land earth via a large, open flameproof switch and throw the switch and connect the ship to land earth. Modern ships may have problems if this is done, so care must be taken to follow company and maritime regulations for this. **WARNING: USING INSULATED HOSES POSES THE RISK OF STATIC CHARGE BUILD UP ON THE HOSE.**

To sum up then, we use hoses with wire coils to stop the build-up of static electricity on the hose material and risk having a static spark to some part of the equipment that has a different potential when loading or unloading powders or liquids.

1) Does your plant have a Jetty?

2) Are all Hoses bonded from end to end?

3) Do all Hoses have a Unique Number?

4) Do you have a routine schedule for Testing these Hoses?

Bonding/Earthing of Large Vacuum Units:

Sometimes the plant may call in large vacuum units mounted on either a trailer or truck to suck powder or liquid out of a vessel. These units usually have a large static arm which hydraulically lifts up and down or swivels. On the end of the static arm there is usually a flexible hose which goes from the arm into the vessel which can then suck out the contents of either powder or liquid, as mentioned above.

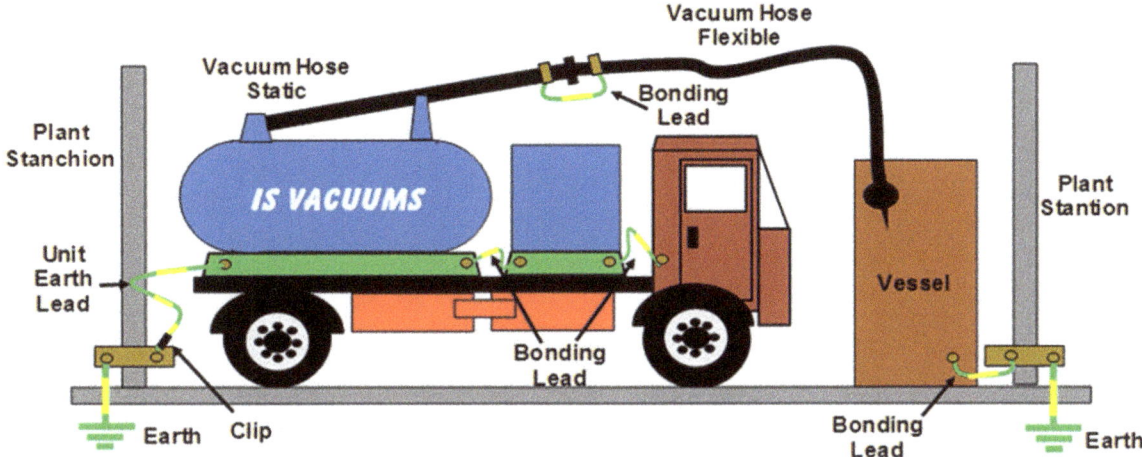

There are several things to look out for in the above example, namely:

1 – The plant may be running whilst this is going on, usually putting the vacuum unit into Hazardous Area Zone 1 or hopefully Zone 2.

2 – The vessel is part of the plant and will have a bonding wire going from its bonding point to a plant earth bar. This will ensure that the vessel is the same potential to earth as the plant structure.

3 – The wagon or trailer should have an earth point somewhere on it where a green and yellow lead is connected so that it can be connected via a clip to the plant earthing system, which would ideally be an earth bar.

4 – The different pieces of equipment on the wagon or trailer should be all bonded together so that when the unit is running they are all of the same potential. Please note that companies may bond their equipment together in all sorts of different ways, not necessarily as shown in my diagram above.

5 – The flexible hose must have a wire coil in the rubber or plastic to ensure that there is the same potential from the tip to where it connects to the static pipe. Care must be taken here that there is a wire coil otherwise the hose, plastic or rubber could charge up to a dangerous potential and cause a spark.

6 – The 'Flexible' hose should be bonded to the 'Static' hose so that any static electricity charge will leak away through the above earthing and bonding measures that have been taken.

Please note that the above diagram and my explanation are only a guide to give an idea of what is happening. This earthing and bonding of the unit can be done in all sorts of different ways. The electrical technician/company must ensure that the unit is safe.

Bonding/Earthing of Road Tankers:

Looking at a chemical factory from outside it is mainly an accumulation of metallic equipment i.e. structures, columns, vessels, tanks etc. Earthing and bonding must be carried out on the complex to ensure that all of this metalwork is at the same potential to reduce the risk of sparking and of course to protect against lightning etc.

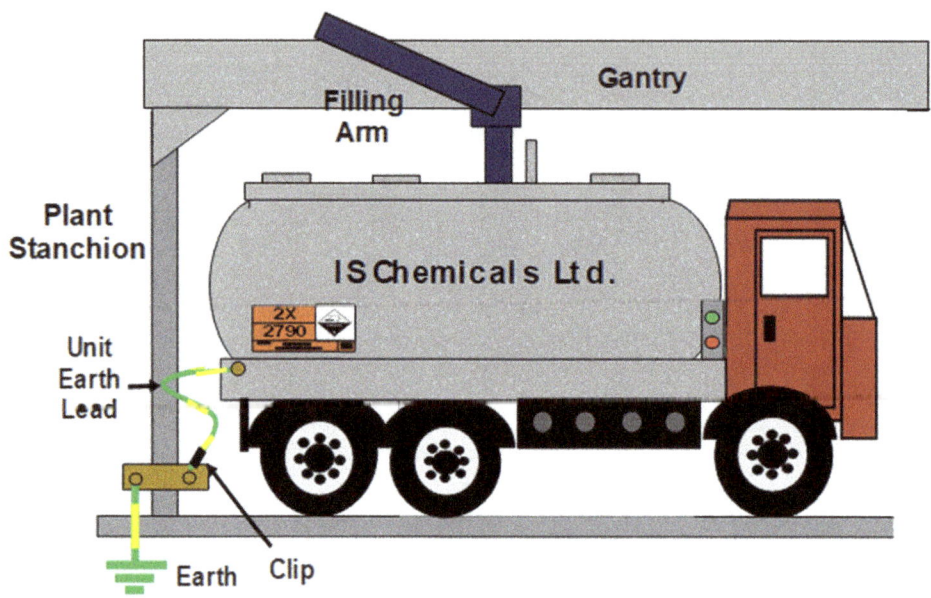

Vehicles are no exception. Usually these days most of the hazardous areas on our chemical complex are Zone 2 (Gases & Vapours) and Zone 22 (Dusts) where we are not expecting the gas, vapour or dust to be present. When the vehicle is being loaded we have to therefore ensure that the metalwork of the vehicle is exactly the same potential as the metalwork around it such as the filling arm, gantry etc. The filling arm will be well bonded to the structure so it will be at earth potential.

We need basically three things on a triangle for combustion and in the case of a vehicle we have namely: **Oxygen** (in the air), **Fuel** (Zone 2 so unlikely) and **Ignition** (vehicle's hot engine or static). So our vehicle can travel through the factory and most of the time there are only two.

When the vehicle is driving through the factory, the likelihood is that only two parts of the combustion triangle are present, namely **Oxygen & Ignition.** The factory roads will most likely be risk assessed as non-hazardous areas so pose no threat and if the vehicle turns off road the chances are it will be driving through a Zone 2 situation so no threat here.

However when the tanker pulls up at the loading gantry and gets ready to be loaded with hazardous chemical, suddenly we can have the third part of the combustion triangle added which is **Fuel.** So now we have a situation where we have all three parts of the combustion triangle, with **Oxygen** (in the air), **Ignition** (in the way of static electricity) also present. We must ensure that, as mentioned above, the metal of the vehicle is at the same potential as the filling arm and gantry and we can do this in two ways; 1) by connecting an earth or bonding lead from the plant metalwork onto the tanker or 2) by using an Earth Monitoring Unit which is discussed on the next page.

Just as a matter of interest, the filling nozzle on the loading arm will go right down to the bottom of the tank so that it is filling into liquid and not 'splash filling' to cut down the risk of a static charge.

Bonding/Earthing of Road Tankers Earth Monitoring:

Looking at a chemical factory from outside it is mainly an accumulation of metallic equipment i.e. structures, columns, vessels, tanks etc. Earthing and bonding must be carried out on the complex to ensure that all of this metalwork is at the same potential to reduce the risk of sparking and of course to protect against lightning etc.

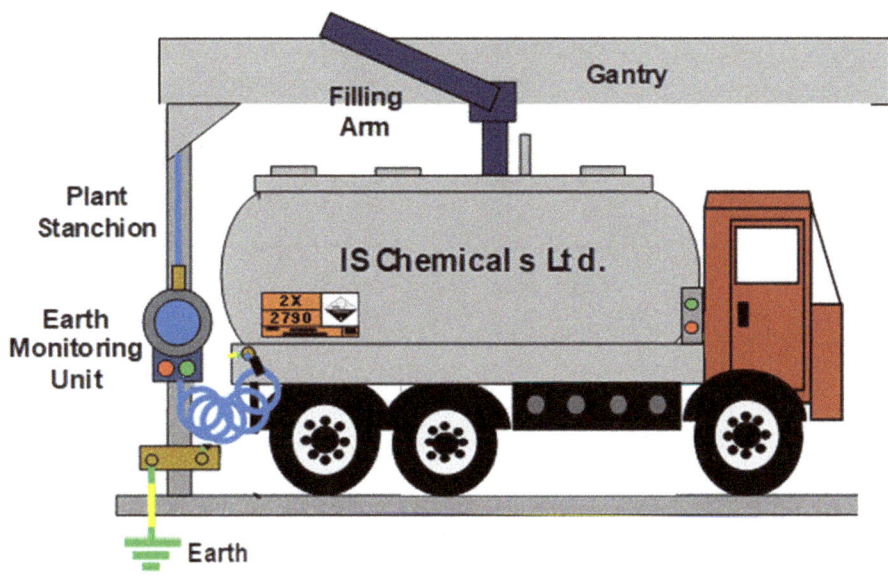

Vehicles are no exception. Usually these days most of the Hazardous Areas on our chemical complex are Zone 2 (Gases & Vapours) and Zone 22 (Dusts) where we are not expecting the gas, vapour or dust to be present. So when the vehicle is being loaded we have to ensure that the metalwork of the vehicle is exactly the same potential as the metalwork around it such as the filling arm, gantry etc. The filling arm will be well bonded to the structure so it will be at earth potential.

We need basically three things on a triangle for combustion and in the case of a vehicle we have namely: **Oxygen** (in the air), **Fuel** (Zone 2 so unlikely) and **Ignition** (vehicle's hot engine or static). So our vehicle can travel through the factory and most of the time there are only two. When the vehicle is driving through the factory the likelihood is that only two parts of the combustion triangle are present namely: **Oxygen and Ignition.** The factory roads will most likely be risk assessed as non-hazardous areas so pose no threat, and if the vehicle turns off road the chances are it will be driving through a Zone 2 situation so no threat here.

However when the tanker pulls up at the loading gantry and gets ready to be loaded with hazardous chemical, suddenly we can have the third part of the combustion triangle added which is **Fuel**. So now we have a situation where we have all three parts of the combustion triangle **Oxygen** (in the air), **Ignition** (in the way of static electricity) present.

Now instead of just clipping a green and yellow wire onto the tanker and hoping for a good connection, or relying on the driver's memory to clip it on at all, a blue coiled wire from a unit called an 'Earth Monitoring system' is connected to the tanker. The lights on the unit change from **red** to **green** when a good connection is formed. No green light, no pumping as this unit is interposed with the pump. Also, if the blue wire is not connected then no pump. The system is intrinsically safe so sparks have no energy to ignite gases, vapours or dusts.

Bonding of Pipe Flanges:

There are certain circumstances where pipe flanges on a pipeline are bonded across the flange using copper tape or braided copper. Remember that these pipes have gaskets in them which are insulators (green in the diagram below) so now we are relying on the stud-bolts making contact with the flange for continuity.

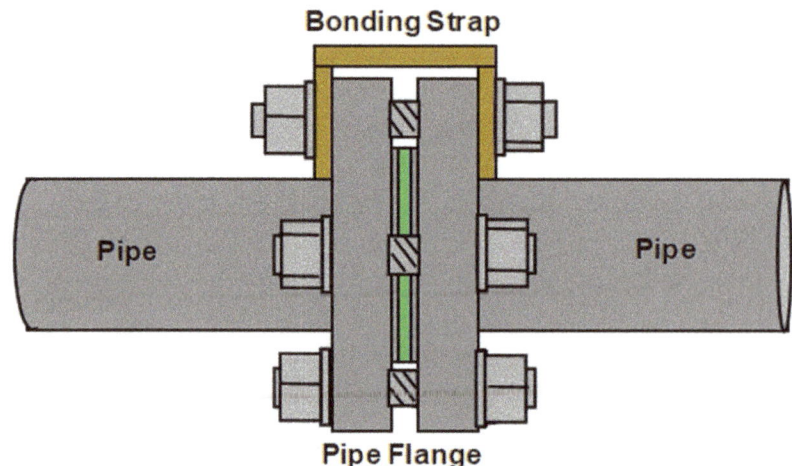

Technically 'rust' is an 'insulator' and therefore will not conduct electricity. However, that is pure rust so the question is do we rely on the stud-bolts which may be rusty to conduct the electricity? The answer is probably yes, but we put on the bonding strap just to be sure that we do not end up with an isolated section of pipe, remembering that the 'gaskets' are also **insulators.** So what can cause static electricity in a pipeline? Firstly we must look at liquid velocity so the greater the velocity the greater the charge. (10m/s is a fairly safe flow.) When filling, we always fill tankers and drums from the bottom up as trickling liquid (splash filling) can cause a static charge in the liquid.

High resistance liquids are the worst as they hold the charge longer. If the liquid ends up in a large storage tank there must be a 'settling' time, sometimes as long as 30 minutes if it is a large diameter tank, to allow the tank earthing system to remove the charge in the liquid.

If the pipe is carrying hazardous liquids, or powder which has a high resistivity and maybe low conductivity, then charge can build up on the material. It must be ensured that any bonding protective measure can take away the charge faster than it can build up. Hydrocarbons are a classic for retaining static charge. Remember petrol is basically made from oil.

These liquids can retain their static charge in pipes and tanks for hours even if the pipes are bonded and the tanks earthed. An additive can be added to the liquid to increase its conductivity, but this may not always be practical or efficient.

High resistivity dust transfer has the same problems as liquids and is sometimes not so easy to remedy. You cannot add conducting agents to dust. All I can say here is that all equipment used for the transfer of dusts must be bonded. All drums must have bonding straps or Earth Monitoring Systems to ensure that there is no build up on the drum.

1) Do you bond across Pipe Flanges?

Bonding of Pipes:

Sometimes **'Earthing Rings'** are used on pipe flanges to earth electrostatic charge in pipelines which may be produced by the liquid that is flowing through them, especially if the liquid is a good insulator. Called 'Earthing Rings', they are actually a bond.

In America these devices may be called **'Earth Spades'**. They may be installed in flanges either side of instruments in the line to control/measure the flow rate such as a flow transmitter would. Earth wires can be taken from the small hole at the end of the stem to each other or plant metalwork. Depending upon the pipe and content material these units may be made of materials to conduct the static such as **'Hastelloy'** metal (nickel, chromium, iron and molybdenum) or **PTFE** containing carbon.

A whole range of electrostatic prevention fittings can be obtained and fixed to the pipes.

So what can cause static electricity in a pipeline? Firstly we must look at liquid velocity so the more velocity, the greater the charge. (10m/s is a fairly safe flow.) When filling, we always fill tankers and drums from the bottom up as trickling liquid (splash filling) can cause a static charge in the liquid. High resistance liquids are the worst as they hold the charge longer. If the liquid ends up in a large storage tank there must be a 'settling' time, sometimes as long as 30 minutes if it is a large diameter tank, to allow the tank earthing system to remove the charge in the liquid.

If the pipe is carrying hazardous liquids, or powder which has a high resistivity and maybe low conductivity, then charge can build up on the material. It must be ensured that any bonding protective measure can take away the charge faster than it can build up. Hydrocarbons are a classic for retaining static charge. Remember petrol is basically made from oil.

These liquids can retain their static charge in pipes and tanks for hours even if the pipes are bonded and the tanks earthed. An additive can be added to the liquid to increase its conductivity, but this may not always be practical or efficient.

High resistivity dust transfer has the same problems as liquids and is sometimes not so easy to remedy. You cannot add conducting agents to dust. All I can say here is that all equipment used for the transfer of dusts must be bonded. All drums must have bonding straps or Earth Monitoring Systems to ensure that there is no build up on the drum.

1) Have you heard of Earth Rings?

Earthing of Pressure Vessels:

When it comes to pressure vessels on the structure they should be **equipotential bonded** and Engineers must carry out a risk assessment as to how much of a problem may be caused by static electricity. As an example I have shown two vessels on a floor of a plant. The vessels are metal, the plant flooring is metal and the stanchions are metal. The vessel numbering system, as far as bonding goes, is irrelevant but vessels on the 2nd floor might be V201 to whatever and 3rd floor V301 to whatever.

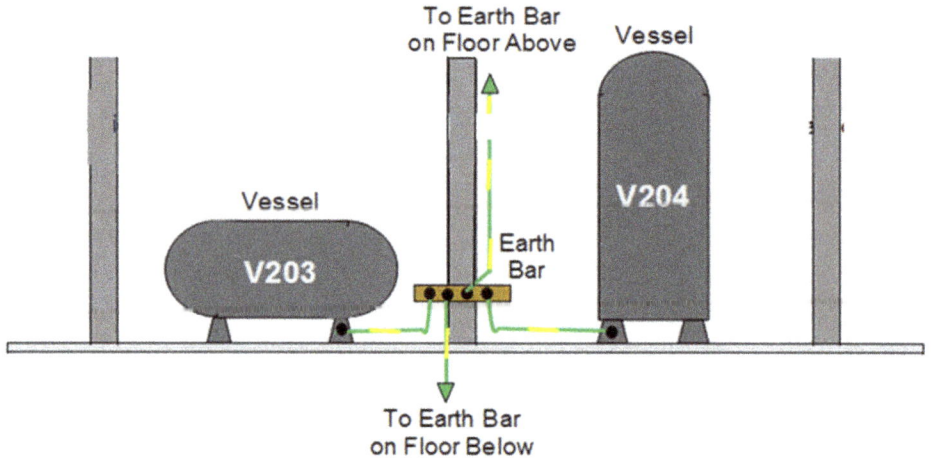

Can we be sure that with everything made out of metal there are good contacts between all of the equipment and the actual structure of the plant? In a risk assessment the answer to the last question would probably be that we cannot be absolutely sure. Remember in the section under pipelines we have gaskets insulating flanges so we have to rely on stud-bolts, and we discussed 'rust' being an insulator of sorts.

These are pressure vessels, not tanks, so the connection to the plant would be an equipotential 'bond' and not an earth although questions are asked such as 'are the vessels "earthed"?' There will usually be only one bonding wire and not two earth wires as on a tank. Now the plant should have several earth rods positioned around the ground floor going to earth bars, probably something like one on each corner of the plant, one in the middle at each side and one, possibly two, in the very centre of the structure. These earth bars will be joined to earth bars on the above floors which at some point will be connected to the metalwork making up the structure of the plant itself, flooring etc.

As mentioned above these are vessels, not tanks, and as such may only require one bond to the structure earth bars. I say connect to earth bars and not the structure itself as these copper bars are interconnected and end up going directly to the plant earth rods and are much easier to connect to. Where the earth point is located on the vessel should be thoroughly clean of paint down to metal and then coated as protection against corrosion.

In the past, individual bonding points could not be tested as they were connected to the plant earthing structure so a good reading would be inevitable. Testing these days can be achieved by using the earth clamp meter (there are several on the market). Putting the clamp meter around the bonding wire will give a reading in ohms of the local earth loop, so we would probably get a reading in the vicinity of 10Ω or better.

Filling Cans with Hazardous Liquid:

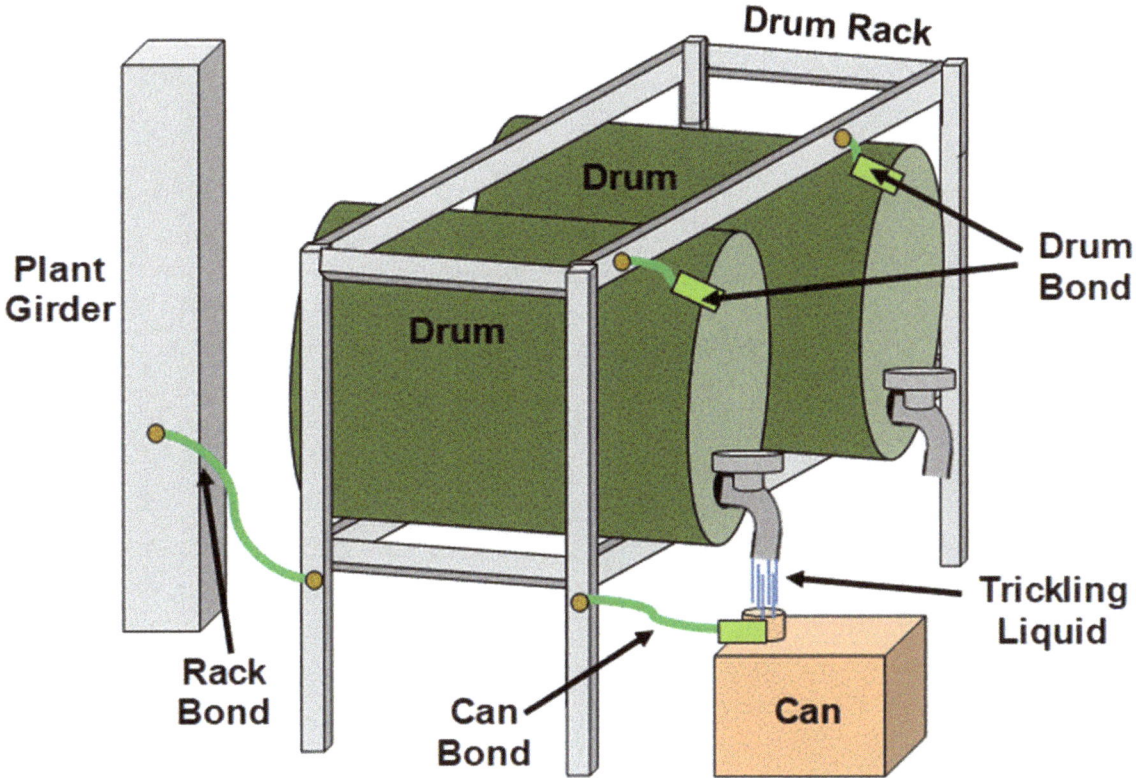

So in this instance we are back to equipotential bonding as well as static bonding in the case of the liquids.

Try to avoid carrying out procedures like the one above. Splash filling a can with trickling liquid is a recipe for trouble. These sorts of procedures carry with them the risk of static sparks. If the use of the drum rack cannot be avoided ensure that all of the metalwork is at the same potential. So above, the rack is bonded to the structure metalwork which could be 10Ω.

Now if you look at the drums above, they are also bonded to the rack. Next the metal container being filled is bonded to the rack. Differences in potential can cause static sparks.

Finally care has to be taken with the liquid. 'Splash' filling can cause the liquid to charge up and if the liquid is not a good conductor this charge may remain for a few seconds until the liquid settles down. By earthing the metal container, the settling down time can speed up enormously and also there are no differences in potential to invite sparks.

Be careful using crocodile clips all over. These are ideal as temporary bonding clips, but try to use clamps and screws which are a little more permanent. Be careful with dissimilar metals used for fixing clips and wires as certain mixed metals will corrode very fast, forming small cells.

1) Have you got any unearthed drum racks on your plant?

2) If YES what is the liquid in the drums?

Bonding of Instrument Junction Box:

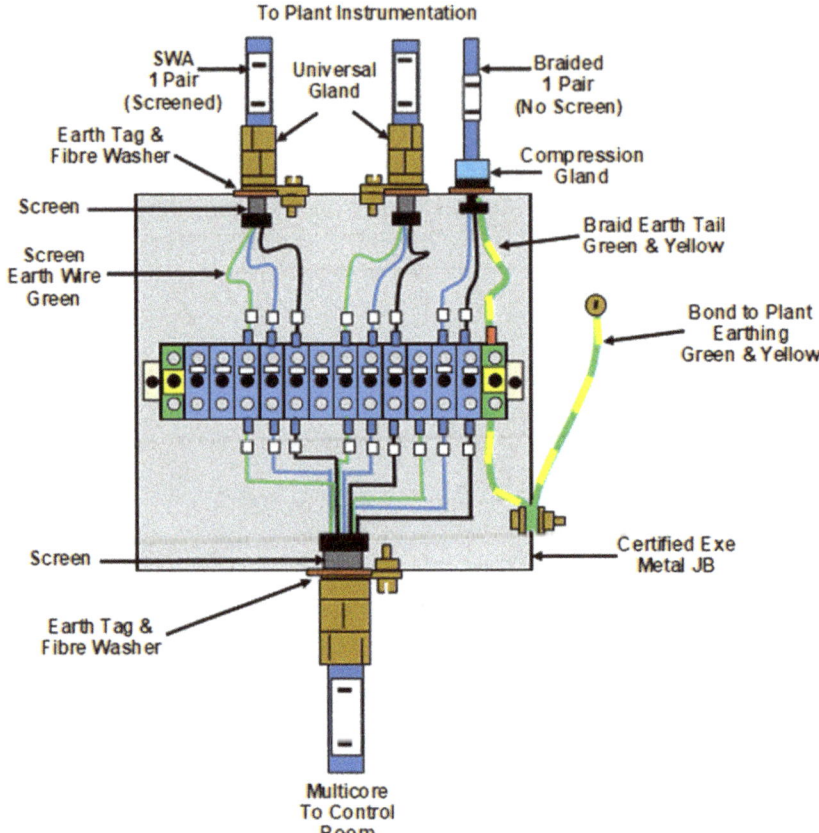

Just to explain a bit about the diagram above, what we have is a typical, intrinsically safe, certified junction box, certified because there is more than one loop in the box. At the base there is an 'IS' multicore cable going back to the control room and at the top there are three 'IS' cables to instruments.

Two of the cables are steel wire armoured (SWA), screened and one is braided. The steel wire armour is bonded to the plant earthing system via the SWA trap inside the universal gland. An earth tag is put onto the gland to ensure a good contact to the pressed steel box. Fibre washers are used where the four glands enter as it is a clearance hole. The fibre washer goes next to the box under the earth tag.

The green screen wire must not be connected to earth at this end so they will be insulated away from earth and via the large multicore they will be earthed, at the barrier end, to the barrier box to wherever the hook up or loop diagram states. Usually this is a **1Ω clean earth**.

The braided cable enters the box via a compression gland and does not have a screen. The braid requires earthing so it is made into a 'tail' inside the box and **green and yellow sleeving** put over and connected to a **5Ω dirty earth** inside the box. **Remember the braid is not a screen.**

Finally the pressed steel box must have a **4mm² minimum** bonding wire from the outside of the pressed steel metal box to the metalwork of the plant. This would apply whether the box is power or IS.

1) Are all of your Instrument Junction Boxes & Marshalling Boxes Earthed?

What is Static Electricity?

Before we discuss Static Bonding we must ask what is static electricity? Just for now let us say that static electricity is so called because it is **static i.e. does not move, as against current electricity, which does move.** Both are to do with electrons orbiting atoms. Let us say that an atom is very similar to a solar system. It has a nucleus which is the sun and particles in orbit which are similar to planets. Everything is made up of atoms and it depends upon the atomic make up as to what that material is. Things might become clearer as we go on.

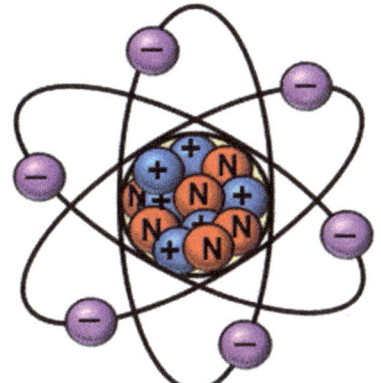

On the left I have shown a diagram of an atom which is made up of three types of particle. In the diagram the nucleus has in it red and blue particles of which the red ones are called '**Neutrons**' and the blue ones '**Protons**'. My atom has six of each. In orbit around the nucleus coloured purple are six '**Electrons**'. This is not the end of the equation though because these particles are made up of even smaller particles called '**Quarks**' which I have mentioned out of interest, but they do not actually enter into our explanation.

Electrons have a '**Negative**' charge, protons have a '**Positive**' charge and neutrons are '**Neutral**' i.e. neither positive nor negative. Neutrons, protons and electrons are usually equal, so a carbon atom (above) has six protons and a hydrogen atom only has one proton. Some materials keep their electrons tightly around the nucleus and will not let them release. With some materials, the electrons are in wide orbits and can easily be 'stolen' or 'borrowed'.

The rubbing of certain materials with others tends to drag electrons from the orbits of the atoms in one material onto the other material. This is called '**Triboelectricity**' or '**Static Electricity**'. Static electricity will only occur on insulators not conductors. You may have carried out experiments at school where a plastic rod is rubbed with a dry cloth and then the rod will pick up bits of paper. (Charged materials will attract opposite charges.)

Materials that will freely give up electrons become more positively charged (leather, glass, nylon, wool etc.) and materials that freely accept electrons become negatively charged. (PVC, polyester, silicon, Teflon etc.) So sit in a car in leather seats wearing polyester trousers and the polyester will gather 'loose' electrons from the leather and you will get out of the vehicle charged up like a capacitor. If you are wearing shoes with insulated soles then you will stay charged until you touch a conductor such as the metal of the car. These triboelectric materials will be waiting to equal their electrons out with a spark.

Chemical plants and platforms insist you wear anti-static safety footwear so that you discharge the moment you step out of the vehicle. Charges of around 10,000V can also build up on your clothing whilst you are walking along especially on a carpet or insulating floor. So where is all this leading to? Well, flexible hoses by their name are usually made of flexible material such as nylon, rubber, plastic, etc. and the liquid, solid material, clothing or in fact air, can cause a charge on the surface which can discharge with a spark. Even isolated metal pipes can charge up if the electrons have nowhere to discharge. This is static electricity so static bonding and earthing must be done to ensure that there cannot be any spark discharges in our Hazardous Areas.

Antistatic Footwear:

So what is antistatic footwear? If you ask most people the question 'What is safety footwear?' they will reply 'safety boots or shoes with steel toecaps' and they would be correct, but there is one other thing required in Hazardous Areas and that is that the soles are made of an antistatic material.

EN ISO 20345

Above is the Standard covering safety footwear and states the contact resistance ranges of antistatic wear including footwear. So what are we looking for? We are looking at the point of contact as the footwear soles touch the floor.

We have mentioned earlier that leather is a material that freely gives electrons and polyester freely accepts electrons so if you are sitting on leather car seats in polyester trousers and are wearing insulated sole shoes, you will get out of the car charged up with triboelectricity (static). If you touch something which is a good conductor i.e. the metal car, you will feel a shock as you discharge. The spark from your discharge may be strong enough to ignite gas, vapour or dust. Now if you got out of the car wearing antistatic footwear you would discharge as you touched the ground.

According to the above Standard, footwear is considered to be antistatic (Conductive) if the contact resistance range is between 100kiloΩ and 1gigaΩ (1kΩ and 1gΩ). Over this value they would be considered to be Insulating and this suitability can be measured with a testing device.

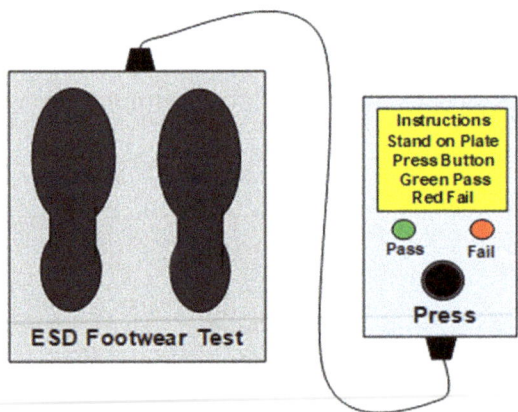

Above I have put a very simplistic diagram of a footwear conductivity tester. The person simply stands on the plate and presses the button on the test unit which would be fixed on the wall in front of them along with any instructions. The test set should be simple with a pass or fail, although more complex testers can be obtained with a screen and scale. It must be noted at this point that ESD stands for 'Electrostatic Discharge' and is different from 'Antistatic', so when footwear is talked about as ESD, this is to another Standard, EN61340. So ESD shoes are antistatic, but antistatic footwear is not always ESD compliant.

1) Do you test your Antistatic Footwear regularly?

Antistatic Wrist Straps:

It is very unlikely that you will be using one of these in your Hazardous Area, but you may use it in the workshop (non-hazardous area). Sometimes going through the way these things work can make antistatic hazards better understood. Please remember that you would probably not be licenced or authorised to repair the electronics of Atex Certified equipment.

Antistatic wrist bands are usually used by people who assemble and repair electronics, especially where there may be CMOS (**C**omplementary **M**etal-**O**xide **S**emi-**C**onductor) technology.

Getting hold of some of these electronics components with fingers would destroy the inside. You can collect a static charge of around 10,000V on your person simply by walking on an insulated floor or wearing insulated footwear. All parts of the inside of the strap should be in contact with your skin for it to be effective.

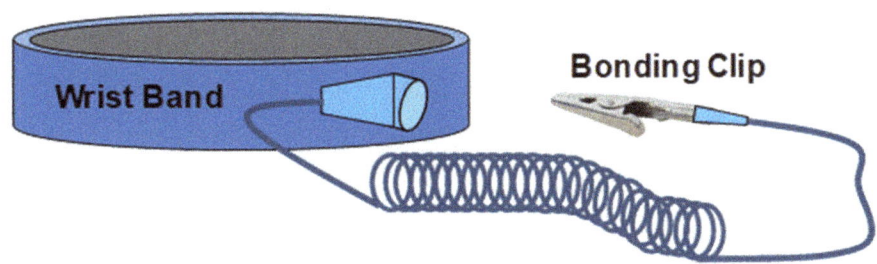

So how can we use this device? Well you can either connect the clip to the metalwork of the equipment that you are working on which would make you the same potential, or you could opt for earthing in which case you would connect the strip to a point that was suitably earthed and this may be your best option (remember our 10,000V). These days, these devices can of course be obtained wireless, but be careful if you are working amongst CMOS (complementary metal-oxide semi-conductor) circuitry as these sometimes can be ineffective.

Remember to test the strap resistance periodically with a strap test device. You should be looking for a resistance of around 4–10Ω. Remember that the resistance doesn't have to be brilliant for electrostatic earthing or bonding so long as it is in the parameters suggested or of course lower. I am sure that I do not have to remind you that in relative terms against live electricity, these wrist bands are a good conductor of electricity so care must be taken that the clip is not clipped onto any equipment where a shock potential is present i.e. Live working.

It is possible to obtain antistatic ankle and foot straps, but these are constrictive and easily forgotten that you have them on! All of the straps can be obtained 'wifi'. These are not new and have been around for a number of years. Just be careful if you go down that route that you obtain all of the available information. I have only ever come across the wired type so I cannot comment.

Antistatic gloves can also be obtained and can be used in conjunction with the wrist band. These gloves can be obtained in a whole range of materials and styles from white cotton to disposable.

1) Have you used Antistatic Wrist Bands or Ankle Bands?

Antistatic Mats:

It is very unlikely that you will be using one of these in your Hazardous Area, but you may use it in the instrument workshop (non-hazardous area). Sometimes going through the way these things work can make antistatic hazards better understood.

Please remember that you would probably not be licenced or authorised to repair the electronics of Atex Certified equipment.

Antistatic mats, more commonly known as 'bench' mats, are generally used by people who assemble and repair electronics, especially where there may be components using CMOS (**C**omplementary **M**etal-**O**xide **S**emi-Conductor) technology for instance.

Taking hold of some of these electronics components whilst charged up with static electricity could severely damage sensitive equipment. You can collect a static charge of around 10,000V on your person simply by walking on an insulated floor or wearing insulated footwear.

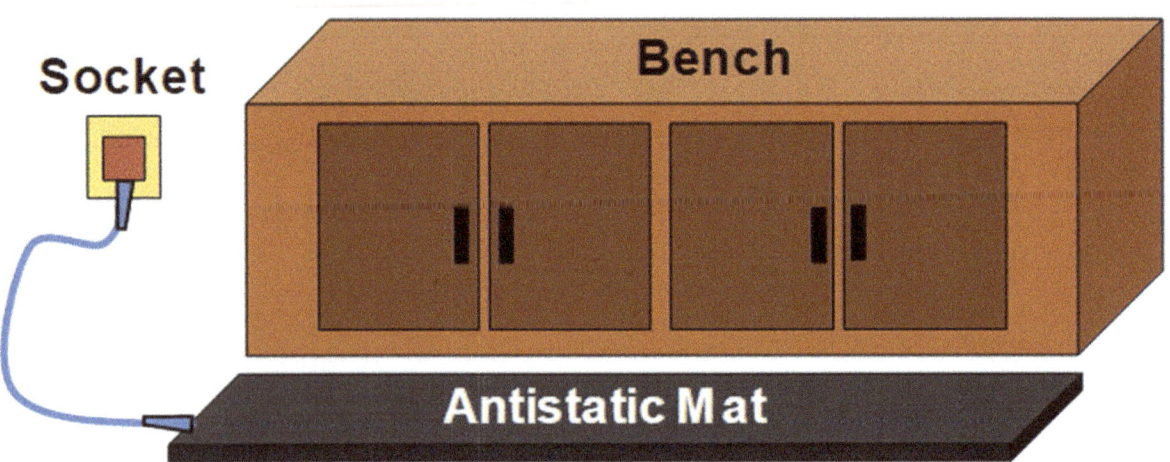

So how can we use this device? Well the mat is an electrical component on its own, and requires plugging into an electrical socket, only using the earth pin of course. (Some manufacturers may advise that the mat be earthed in some other way. Smaller antistatic mats can be used under equipment.)

The mat itself is made from a conductive material that actually collects the static charge and dissipates it through the earth that is provided through the flex. Ideally you will also be wearing antistatic footwear where the soles are also antistatic.

Usually by design the mat is made of two layers of different material connected together. There is a top layer composed mainly of antistatic rubber, which actually collects the static from people or equipment placed on it, and a connected carbon-based layer at the bottom and this is what the earthed cord is connected into.

Care must be taken to ensure that the mat is constantly earthed by plugging into a socket or whatever the manufacturers have designed. Failure to earth these mats could in actual fact have the opposite effect than what it is designed for as the mat could charge up with static leaving the electrons nowhere to go.

1) Do you work on Electronic Circuits where this mat may be used?

Antistatic Spray:

We can spray **CERTAIN** materials with an antistatic spray to decrease the static electricity that the material can collect. It can be obtained in 'mister' form, like my diagram below left, or in 'aerosol' form below right. The beauty of the 'mister' below left is that you can actually make your own antistatic liquid.

Antistatic sprays work by actually making the sprayed material more conductive by using a conductive polymer which would allow the 'borrowed' electrons to flow away more freely. We mentioned in other sections that a person can build up a charge of up to 10,000V on their body. You can sometimes hear the crackle when you remove your sweater.

If, say, in an Instrument Department, technicians work on electronics or very delicate instrumentation, then static electricity at more than 20V can do harm let alone the 10,000V on surfaces. Remember static electricity, even at low voltage, is not intrinsically safe and can cause explosions.

What is the antistatic liquid composed of? Well firstly, to make it bulk liquid, the main ingredient is water. Not ordinary water but de-ionised water that is used in large factory boilers. This is getting on the lines of pure water. Ordinary water contains ions (which can carry electric charge) from copper or iron in pipes or certain ground minerals. These ions have to be removed so if we pass the water through a resin filter, the ions in the water match ions in the resin and go out of the water. The de-ionised water is then mixed with an alcohol solution.

When the final solution is sprayed onto a material, there is a film-like form left on the material which is conductive so static will not build up. Be careful using outside as the rain may mix and form a solution that can have a detrimental effect.

The spray can be used on many surfaces. Ensure that you read the information that comes with the spray. Be extremely careful spraying directly at equipment such as monitors. Ensure that the spray content is suitable.

Protection:

Electrical equipment on a chemical complex requires several different types of protection to prevent gases, vapours and dusts being ignited by arcs and sparks that may be produced by equipment mechanisms. These items of equipment require different earthing methods to allow them to be located in certain areas. Let us look at some of the different types of protection:

1 – Flameproof (Exd): This type of equipment is usually quite bulky and can withstand explosions from within. Any equipment containing a switch will have sparks out in the open inside the equipment. Flanged equipment will not allow explosions to propagate from inside to the outside so will have flameproof paths and have gaps that must be checked with feeler gauges after all maintenance. Other types, besides flanged, are spigot and screw and can be installed in **Zones 1 and 2**. Exd also covers some types of mechanical equipment. As this equipment is made of metal there is usually a bonding screw somewhere on the equipment to connect a **4mm^2** bonding wire to the structure.

2 – Increased Safety (Exe): This type of equipment is usually made of plastic, although can be light metal, so cannot withstand explosions from within. No sparks out in the open here as they are usually inside small certified chambers, so no flanges to check, only seals to IP54 minimum. Can be installed in **Zones 1 and 2**. Can contain an earthing plate inside so that glands are equipotential. As this equipment is made of plastic it is no use putting a stud on it for a bonding wire, so some other method of bonding must be adapted. Usually, the way this is done is to put an earth tag on one of the glands so that a **4mm^2** bonding wire can be attached to the structure bonding all of the glands and the earth plate inside. Note: if there is no earth plate then each gland must have an earth tag.

3 – Intrinsic Safety (Exi): This equipment is usually used on instrumentation and has extra interface units in the loops (Instrument Circuits) such as Zener or Galvanic barriers to ensure safety. Arcs and sparks are very low intensity and will not ignite gases, vapours or dusts, even live with the lids off. Subdivisions Exia, Exib & Exic ensure that the equipment is manufactured safe, even with numbers of faults on them. Exia equipment is safe with two faults and can be installed in **Zones 0, 1 and 2**. Exib is safe with one fault and can be installed in **Zones 1 and 2**. Exic (nL) is just a piece of equipment and faults do not enter into the equation so can be installed in **Zone 2 areas ONLY**. Bonding is achieved by a **4mm^2** bonding wire from earth tags if there is not a bonding screw on the actual equipment. **Manufacturer's documentation must be followed to the letter with this equipment. Company must have accurate Loop Diagrams.**

4 – Reduced Risk (Exn): Usually made out of plastic, but can be light metal. Very similar to 'Exe' but not quite so highly certified. Restricted Breathing comes under this protection. Must be installed in **Zone 2 areas ONLY**. Earthing is usually achieved as in Exe by a **4mm^2** bonding wire from earth tags.

5 – Encapsulated (Exm): This equipment is completely sealed. Subdivisions are Exma, Exmb and Exmc. Exma can go in **Zones 0, 1 and 2**, Exmb can go in **Zones 1 and 2** and Exmc can be installed in **Zone 2 areas ONLY**. Again bonding is achieved by a **4mm^2** bonding wire from earth tags if no bonding screws. **Manufacturer's documentation must be followed to the letter with this equipment.**

6 – Pressurised (Exp): This type of equipment is pressurised with an inert gas, usually air, to a positive pressure and is designed to keep the gas, vapour or dust out. The equipment is usually sited in a metal cabinet of some sort or a special piece of certified equipment. This equipment can be installed in a **Zone 1 or 2** area. Exp now also covers some mechanical Atex equipment of this type of protection. Equipment Bonding again is by a **4mm^2** bonding wire from earth tags if there are no bonding screws on the equipment. If this is a piece of certified equipment there will usually be a bonding screw. If not we are back to earth tags.

7 – Oil immersed (Exo): This type of protection is mainly transformers and is intended to smother arcs or sparks under a medium of oil. Bonding points are usually provided to connect by **4mm²** wire to the plant. Protection can vary and some equipment can go in **Zone 1 and 2. Manufacturer's documentation to be followed at all times.**

8 – Powder/Quartz Filled (Exq): This equipment used to be called quartz filled as possibly originally it was filled with sand. These days it is usually filled with small glass or silica balls which are intended to smother arcs or sparks under a medium. Although the manuals state that this protection can go in **Zones 1 and 2,** it is usually reserved for equipment such as chokes in certified fittings. **All I can say about bonding is that it is as per manufacturer's documentation.**

9 – Special Protection (Exs): Special protection is usually when the Notified Body which certifies the equipment cannot decide which protection it actually fits into so they give it a 'special' label and a lot of documentation as to where it can actually go. Sub divisions Exsa, Exsb and Exsc. Usually on equipment like gas detectors. **All I can say here is that it is bonded as per manufacturer's documentation.**

10 – By Enclosure (Ext): This is a protection that only applies to Dust Zones and not gases & vapours. Mainly concerned with keeping temperatures down and sealing of equipment. Subdivisions Exta, Extb & Extc. Exta **Zones 20, 21 and 22,** Extb – **Zones 21 and 22** and Extc **Zone 22 ONLY.** Bonding is by **4mm²** from a metal screw on the equipment or earth tags. **Manufacturer's documentation to be followed at all times.**

11 – Mechanical (Exh): Exh is a ISO80079 protection that only applies to Atex mechanical equipment encompassing c – Constructional Safety, k – Liquid Immersion and b – Protection by Monitoring. (See my book 'Hazardous Areas for Technicians'.) **There are no electrical components here so mainly interested in equipotential bonding.**

Hazardous Areas:

Zones, Categories & Equipment Protection Levels (EPLS):

We are still working to **EN60079 (Atex)** with **IEC60079** as the incoming Standard so the two Standards are running in parallel at the moment. Dusts have gone from **BS61241** into **IEC60079 – 10/2** with Gases & Vapours being **IEC60079 – 10/1**.

Hazardous Areas are usually divided into Zones. Zone 0, 1 and 2 for Gas and Vapour Areas, and for Dust Zones just put a '2' in front to make Zone 20, 21 and 22.

Zone 0 (Gases & Vapours) & Zone 20 (Dusts):

Hazardous Atmosphere Constant or for long periods of time.

Zone 1 (Gases & Vapours) & Zone 21 (Dusts)

Hazardous Atmosphere likely in normal operation.

Zone 2 (Gases & Vapours) & Zone 22 (Dusts)

Hazardous Atmosphere unlikely or for short periods of time.

Zones & Categories Gases & Vapours:

If we take the Zones, Atex Categories of equipment would be as follows:

Zone 0 – The Category of equipment would be **1G**
Zone 1 – The Category of equipment would be **2G**
Zone 2 – The Category of equipment would be **3G**

> Gases & Vapours Categories are explained in more detail in my book: 'Hazardous Areas for Technicians'

Zones & Categories Dusts:

Zone 20 – The Category of equipment would be **1D**
Zone 21 – The Category of equipment would be **2D**
Zone 22 – The Category of equipment would be **3D**

> Dust Categories are explained in more detail in my book: 'Hazardous Areas for Technicians'

Zones & Equipment Protection Levels Gases & Vapours:

Zone 0 – The Equipment Protection Level is **Ga**
Zone 1 – The Equipment Protection Level is **Gb**
Zone 2 – The Equipment Protection Level is **Gc**

> Gases & Vapours Equipment Protection Levels are explained in more detail in my book: 'Hazardous Areas for Technicians'

Zones & Equipment Protection Levels Dusts:

Zone 20 – The Equipment Protection Level is **Da**
Zone 21 – The Equipment Protection Level is **Db**
Zone 22 – The Equipment Protection Level is **Dc**

> Dust Equipment Protection Levels are explained in more detail in my b\ook: 'Hazardous Areas for Technicians'

Gas Groups:

When looking at Hazardous Areas there are certain parameters that must be correct. These are the Zone, Gas Group, Temperature Class and Protection. The electrical and instrument equipment must also be Atex of course. In the mechanical world pumps, compressors etc. of the future might have to be Atex. Now that we have left the EU there is no knowing what markings will stay. We have discussed Zones and Protection on the previous page so let us now look at Gas groups.

Gas Group:

Gases & Vapours are divided into three Gas Groups namely **IIA, IIB** and **IIC**. It is very important when selecting your equipment that you know what atmosphere it is going into and in whatever Zone that atmosphere is determined to be.

Gases and vapours are divided into Gas Groups based on parameters such as their **Maximum Ignition Energy** and their **Maximum Experimental Safe Gap (M.E.S.G.)** which is explained in more detail in my book: **Hazardous Areas for Technicians.**

IIA Gases/materials include: Ammonia, Benzene, Butane, Cyclohexane, Ethanol, Formic Acid, Heptane, Hexane, Kerosine, Methane (Industrial), Octane, Phenol, Propane, Styrene, Xylene. (There are many more in the tables.)

There are many more well-known gases/materials that are IIA than there are IIB.

IIB Gases/Materials include: Diethyl Ether, Ethyl Methyl Ether, Ethylene, Formaldehyde, Hydrogen Cyanide, Hydrogen Sulphide. (There are a few more in the tables.)

IIA Gases/Materials come out top in numbers, well ahead of IIB, but there are only three IIC Gases.

IIC Gases: Hydrogen, Acetylene and Carbon Disulphide.

Looking at the above Gas Groups, **IIC** is the worst requiring only around **20μJ** of energy to ignite whereas Gas Group **IIA** would require around **200μJ** of energy and **IIB** around **110μJs** to ignite. So **IIC** equipment can be put into a **IIA** and **IIB** Gas environment.

IIB Gas Group, as above, requires around **110μJ** of energy to ignite so this can be put into a **IIA** Gas environment.

So now the Gas Group, IIA, IIB or IIC, must be obtained for **EVERY** Gas/Vapour/Material on the plant where the equipment is to be installed. Then the Gas Group of the equipment must be chosen based on the worst case scenario i.e. IIC followed by IIB followed by IIA.

Obviously cost is involved here as the IIC equipment may have many more special conditions on installation than IIB or IIA equipment. The sealing might be more enhanced on say Exe Increased Safety Equipment to, say, IP66.

Many Electrical Engineers choose the IIC option for safety and also because the equipment could be fitted into **ANY** Gas Group II. We now go on to the **Temperature Classification** on the next page.

Dust Groups:

We have had a look at Zones, Protection, Gas Groups and Temperature Classes in the Gases & Vapours world so now let us look at the Dust world.

When the Zone is a Dust Zone and not Gases & Vapours there is not a great deal of change. You just have to remember to put a '2' in front making them Zone 20, Zone 21 and Zone 22.

The rules still apply i.e. Continuous or for long periods of time, Likely in normal operation, or Unlikely or for short periods of time.

In the Gases & Vapours world the Gas Groups are IIA, IIB & IIC. Well, in the Dust world the Dust Groups are as follows:-

IIIA – Combustible Flyings which might be materials like Kapok, Jute etc.

IIIB – Non-Conductive Dust such as Wood which of course is still Combustible.

IIIC – Conductive Dust such as Coal, Carbon etc.

Just like Gases & Vapours, IIIC is the worst and can go into IIIB & IIIA.

Again as in Gases & Vapours you need to know what the atmosphere will consist of in the area where you are installing the equipment.

Earthing requirements do not change a lot because the medium is dust.

Temperature Classification:

When choosing equipment for a Hazardous Area there are certain parameters that must be correct: the Zone, Gas Group, Temperature Class and Protection. The equipment must also be Atex of course. We have discussed Zones and Protection and Gas Groups on the previous pages so let us now look at Temperature Classification.

Temperature Classification Gases & Vapours:

We have looked at the Zones and Gas Groups where our equipment is going to be installed, now let us look at what Temperature Classification our equipment is going to be. The manufacturers will state that the Temperature Class of the equipment (Green) will not go over the temperature stated (yellow) **under normal or specified fault conditions.** It does not say that the temperature of the motor will **NEVER** go over the temperature in yellow.

T. Class:	Temperature:
T1	450°C
T2	300°C
T3	200°C
T4	135°C
T5	100°C
T6	85°C

In this country we have six Temperature Classifications, namely T1 – T6 and they have Maximum Surface Temperature next to them, 85 Degrees Celsius to 450 Degrees Celsius. The Temperatures in yellow are quite high, for instance you will burn your hands if the Temperature reached 85 Degrees Celsius and it would melt lead at around 300 Degrees Celsius!

The figure in yellow must **NOT BE HIGHER** than the gas/vapour Ignition Temperature where it is installed. otherwise the equipment could ignite the gas if it did reach its T. Class.

Some examples of Temperature Classes are below. So the Temperature Classification (yellow above) must be **LOWER** than the Gas/Vapour/Dust Ignition Temperature.

Again, Electrical Engineers tend to go for T6 for safety so our safest equipment gas group and temperature classification would be:

IIC T6

Temperature Classification Dusts:

With Dust equipment you will see an actual temperature on the equipment like T80C. This is the maximum surface temperature that the equipment will reach with a ballpark figure of a **5mm** layer of dust. If the equipment dust layer is expected to be greater than **5mm** then the manufacturers should be contacted. There will also be a Temperature Class on the equipment which will be around the surfact temperature stated.

So you find out what dust(s) you are dealing with and find its Layer and Cloud Ignition Temperature(s) which will be different. You then work out the figures below for those dusts and pick the worst case scenario. (The one with the **lowest** Ignition Temperature.)

The equipment Temperature Class must be 75 degrees below the dust **Layer** Ignition Temperature and 2/3rds (66%) of the **Cloud** Ignition Temperature.

Several Earthing Terminologies:

1 – **Earth:** This is the medium of earth surrounding the electrode and determines the resistance and path back to the transformer star point. Often referred to as Terre.

2 – **Earthing:** Metal objects are earthed directly to earth rods. This could include plant earth bars.

3 – **Grounding:** This is the North American term for Earthing.

4 – **Bonding:** This is the action of joining two metal objects together to achieve, say, equipotential bonding, but no connections directly to earth.

5 – **Earth Electrode:** This is of course the main part of the Earthing System. This can take the form of Rods, Pipes, Lattice, Plates etc.

6 – **Earth Lead:** The wire, usually about 70mm from the Earth Electrode to, say, the Earth Bar.

7 – **Earth Pit:** The hole dug to house the Earthing electrode. Sometimes filled around the electrode with earthing Backfill Compound.

8 – **Earth Pot:** The protective box on top of the Earthing Electrode which houses the Earth Cable connection. Can be plastic or concrete. (On top of the Earth Pit.)

9 – **Earth Resistance:** Usually in ohms (Ω) and is the resistance between the Earth Electrode and the ground

10 – **Exposed Conductive Part:** A metal part which could become live under fault conditions, say a metal cover of electrical equipment.

11 – **Extraneous Conductive Part:** A metal part not part of the electrical equipment such as metal cable tray, conduit, plant structure which requires equipotential bonding.

12 – **Equipotential Bonding:** The connecting together of exposed conductive parts and extraneous conductive parts to ensure equal potential so less chance of fatal shocks.

13 – **Step Voltage:** Estimated by taking a person with a stride of 1 metre and it is what the voltage would be from foot to foot.

14 – **PEN Conductor:** Protective Earthing Conductor. A conductor providing the functions of both Earth and Neutral Conductor i.e. TNC-S System.

15 – **PME System:** Protective Multiple Earthing. The Neutral is earthed in several places from the consumer back to the distribution transformer.

16 – **Main Bonding:** Similar to direct earthing this is where we take a wire from an earth bar straight to, say, a vessel.

17 – **Supplementary Bonding:** This is where we take a wire, say, from one vessel to another.

18 – **Measurement Ω/M:** Many ground potentials are measured in Ohms/Metre. So if I take 1 metre of the ground medium what would be the resistance in Ω.

19 – **Prospective Fault Current (PFC):** The maximum current that will flow in the event of a fault to earth. (See BS7671 – 18th Edition)

20 – **Prospective Short Circuit Current (PSC):** The maximum current that will flow in the event of a short circuit between conductors. (See BS7671 – 18th Edition.)

INDEX

Antistatic Safety Equipment	110–13
Atex Equipment Earthing	98–99
Auto Transformer	76
Bentonite Compound	36
Bonding	95, 120
Bonding Atex Equipment	98
Bonding Drum Racks	107
Bonding Pipe Flanges	104
Bonding Pipework	105
Bonding Road Tankers	102
Bonding Ships' Hoses	100
Bonding Vacuum Units	101
Busbar Earthing	68
Bus-Zone Protection	65
Cable Management'	14
Cadweld Jointing	33
Centre Tap Transformer	72
Class I	16
Class II	90, 91
Class III	74
Clean Earthing	55–57
Common Mode Voltage	85
Conduit Earthing	35
Converters	85
Core Balance for Motors	53
Current Operated ELU	50, 51
Current Transformer	77
Delta Connected Windings	67
Distribution Transformer	10, 12, 71
Double Insulation (Class ll)	90
Dust Groups	118
Earth Bars	26
Earth Cable Colours	11
Earth Cables	35
Earth Clamp Meter	42
Earth Connections	33
Earth Description	10, 120
Earth Electrode Accessories	32
Earth Electrode Coil	31
Earth Electrode Compound	36–37
Earth Electrode Concrete	30
Earth Electrode Electrolytic	29
Earth Electrode Gel	36–37
Earth Electrode Lattice	28
Earth Electrode Plates	27
Earth Electrode Resistance	38–41
Earth Electrode Ring	61
Earth Electrode Rods	25, 26
Earth Fault	15, 16, 62
Earth Leakage Unit (ELU)	47–49
Earth Loop Impedance (TN-S)	46
Earth Loop Impedance (TT):	47
Earth Mat/Grid	27, 28
Earth Monitoring	103
Earth Path	15, 16
Earth Pits	25, 32, 34, 120
Earth Pots	25, 32, 34, 120
Earth Shells	24, 25
Earthing Drawing Symbols	89
Earthing Methods	24
Earthing Pressure Vessels	106
Earthing Systems	11, 17, 120
Earthing Transformers	13, 67
Electric Discharge Machining	85

Electric Shock	14, 15
Equipotential Bonding	14, 96–97, 106, 120
Exposed Conductive Part	14, 15, 96–97, 120
Extraneous Conductive Part	14, 15, 96–97, 120
Flameproof (Exd)	114
Floating Roof Tank	84
Fluting	88
Frosting	88
Galvanic Barrier Earthing	56
Gas Groups	117
Generator Earthing	78–81
Ground Resistance Readings	40
Ground Resistance Testing	38–42
Grounding	10, 120
High Voltage Earthing	63–69
Increased Safety	114
Instrument Transformer	73
Intrinsic Safety	114
Inverter	85–88
Isolation/Floating supply	74, 79
IT Earthing System	17, 23
Lightning Protection	58–61
Lock-out Box	68, 69
Main Bonding	95, 120
Marconite Compound	28, 36
MCB–B–C–D Earthing	46–47
Mechanical Equipment	115
Metal Equipment (Class 1)	15–16
Metrosil	62
Motor Earth Path	48
Ohms/Metre	40, 120
PEN Conductor	18, 19, 120
Petersen Coil	24
Pipe Earthing Rings/Spades	105
Pressurisation	114
Prospective Earth Fault	46, 47
Pulse Width Modulated	85
Re-enforced Insulation	91
Residual Current Device (RCD)	50–51
Restricted Earth Fault	63, 66
Rust & Earthing	104
Screen Colours	11
Screens	55–56
Socket Earth Polarity Tester	94
Star Point Earth	12–23
Star Point Reactance Earth	23
Star Point Resistance Earth	23
Static Electricity Basics	109
Steaming Out Bay	93
Step Up Transformer	75
Step Voltage	38, 63, 64, 120
Storage Tank Earthing	82–84
Supplementary Bonding	95, 120
Surge Protection	62
Temperature Classification	119
TNC Earthing System	20
TNC-S Earthing System	18
TN-S Earthing System	21, 46
Transformer Delta Delta	67
Transformer Delta Star	10, 67
Transformer Star Delta	67
Transformer Star Star	67
Triboelectricity	109
TT Earthing System	17, 22, 47
Variable Frequency Drives	85–88
Variac	76
Varistor	62
VFD Rotor-Shaft Current	87
VFD Stator-Ground Current	88
VFD Stator-Rotor Circ. Current	86
VFD Stator-Shaft Current	87
Voltage Operated ELU	52
Water Jetting/Cutting	92
Zener Barrier Earthing	55
Zones	116

Lightning Source UK Ltd.
Milton Keynes UK
UKHW051141060223
416527UK00007B/180